Chicago

Center Books on Chicago and Environs

George F. Thompson,

series founder and director

Chicago A GEOGRAPHY OF THE CITY AND ITS REGION

BY JOHN C. HUDSON

THE CENTER FOR AMERICAN PLACES
SANTA FE AND STAUNTON

THE UNIVERSITY OF CHICAGO PRESS
CHICAGO AND LONDON

PUBLISHERS' NOTES: *Chicago: A Geography of the City and Its Region* is the tenth volume in the series *Center Books on Chicago and Environs*, George F. Thompson, series founder and director. The book was brought to publication by the Center for American Places and the University of Chicago Press in an edition of 2,250 hardcover copies, with the generous support of the Graham Foundation for Advanced Studies in the Fine Arts and the Friends of the Center for American Places, for which the publishers are most grateful. A limited edition of 250 copies of *Chicago Portfolio: Where Geography and Photography Meet* (ISBN: 1-930066-45-7) was published separately by the Center for American Places in commemoration of the annual meetings of the Association of American Geographers and the Society for Photographic Education, held in Chicago in March 2006. For more information about the Center for American Places and the publication of *Chicago: A Geography of the City and Its Region*, please see page 260.

The Center for American Places, Inc.
P.O. Box 23225
Santa Fe, New Mexico 87502, U.S.A.
www.americanplaces.org

The University of Chicago Press, Chicago 60637
The University of Chicago Press, Ltd., London
www.press.uchicago.edu

15 14 13 12 11 10 09 08 07 06   1 2 3 4 5

Library of Congress Cataloging-in-Publication Data is available from the publishers upon request.

ISBN 0-226-35806-2 (hardcover)

Frontispiece: Millennium Park, 2004. Photograph by Brook E. Collins. Used by permission of the artist and the Chicago Park District.

To Hudson, my grandson

CONTENTS

Part III. Historical Patterns

Part IV. The Growing City

Part V. The Changing City

LIST OF FIGURES

PREFACE AND ACKNOWLEDGMENTS

Credit for the idea behind this book goes to George F. Thompson, founder, president, and publisher of the Center for American Places. Several years ago, in conversation, he noted that a new geography of Chicago was long overdue. Indeed, it has been more than thirty years since the first edition of Irving Cutler's *Chicago: Metropolis of the Mid-Continent* was published. Cutler's book remains a remarkable and usable synthesis of the subject, but much has changed in the years since it appeared. George also took on the task of curating a gallery of photographs of Chicago places and people to provide a visual tribute to a city where, as George notes, "geography and photography meet."

Chicago was once the city most studied by American geographers. The emergence of urban geography as a field of study drew from the works of Charles C. Colby, Chauncy D. Harris, Edward L. Ullman, Malcolm J. Proudfoot, Harold M. Mayer, and Brian J. L. Berry, all of whom wrote from academic homes in the Chicago area. As the subject grew in scope in later years, other Chicago geographers, including Michael Conzen and Pierre DeVisé, added new dimensions to our understanding of Chicago.

The production of new geographical studies of Chicago and its region slowed noticeably after the mid-1980s. Loss of departmental status for geography at the three leading universities of the area resulted in a sudden drop in theses and dissertations with Chicago themes, and, shortly thereafter, to a predictable scarcity of scholarly studies appearing in print. Chicago itself declined in size and dropped in population rank from second to third, behind Los Angeles and New York City. The most influential book on Chicago during this period, William Cronon's *Nature's Metropolis*, refocused Chicago history away from the city by projecting it onto the broader scale of environmental history.

In the meantime, innovative urban geography migrated from Chicago to Los Angeles. A revival of city-focused geography emerged during the 1990s in the works of William A. V. Clark, Michael Dear, Allen Scott, Edward Soja, and other Los Angeles-based scholars. Although their work took various ideological perspectives, it was the city as an entity that engaged their interest. Production of a new geography of Chicago, emphasizing changing geographical patterns and recent trends, is a continuation of this revival.

It was decided at the outset that this book would follow past traditions of geographical scholarship on Chicago: it would be data-based insofar as possible, and it would rely heavily on statistics published in the United States Census that are available to anyone. Work on the maps began as soon as detailed census tract tables were available online for the 2000 Census of Population. I am extremely grateful to my research assistant, Tiffany Grobelski, who spent twelve months downloading census data and entering it in the mapping program (AtlasGIS) used to produce all of the maps in this book. Published data from past censuses were converted to the same statistical base to produce maps for past time periods.

Amber K. Lautigar and Rebecca A. Marks of the Center for American Places expedited production and provided friendly support. Kristine K. Harmon, of Charlottesville, Virginia, was a very able copyeditor, and David Skolkin, of Santa Fe, New Mexico, was skillful in his book design. Two anonymous reviewers made useful suggestions for improving both text and maps. And Christie Henry helped to make copublication possible between the University of Chicago Press and the Center for American Places.

My wife, Debby, my children, and my grandchildren supported me and urged me on to the finish. The book is dedicated to one of my grandchildren, Hudson Robert Ridley, who was born midway through the book's preparation. To all in my family go my love and appreciation for their support

Part I

INTRODUCTION

Chapter 1
CHICAGO AND ITS REGION

LIKE ALL CITIES, CHICAGO is a well-defined entity that has legally specified bound-aries that can be shown on a map. Chicago's "region" is more difficult to define. It is a concept understood in general terms, finding expression in labels such as "Chicagoland," popularized by the *Chicago Tribune* and the Chicago Motor Club, but there are no formal boundaries of such a region. The lack of a single definition makes the point that regions are invented for specific purposes and hence they vary in size and shape depending on the use that is intended. The U.S. Census defines urban regions in a consistent manner for purposes of collecting and publishing data about them. There are also functional regions, such as trade areas, that encompass much more territory and which are useful in understanding the city's role in a broader economic context.

THE CENSUS MODEL

The U.S. Census defines several levels of regional boundaries for large cities such as Chicago (Fig. 1.1). The city of Chicago (population 2,896,016 in 2000) is the most basic level. The city's familiar shape, hugging the shoreline of Lake Michigan, acquired an outlier in 1958 when O'Hare Airport and the lands bordering it were annexed. Apart from that addition, Chicago's boundaries have changed very little for many years.

The Chicago urbanized area (population 8,307,904 in 2000) is the next higher level in terms of Census definition. An urbanized area consists of a central city of at least 50,000 people plus all adjacent areas having a population density of at least 1,000 per-sons per square mile. Chicago's urbanized area extends for some distance west into DuPage County, northward across Lake County to the Wisconsin border, and east-

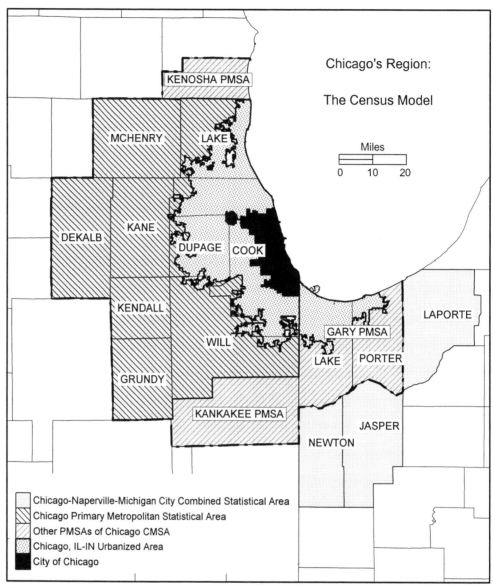

Fig. 1.1

ward into Lake and Porter counties, Indiana. Urbanized areas are the most precise definitions of urban regions in terms of settlement density, although they do not include all territory where daily commuters to the central city reside.

The most commonly used regional-level statistical unit for large cities is the Metropolitan Statistical Area (MSA). An MSA consists of a central city of at least 50,000 people, the remainder of the county in which it is located, plus adjacent counties that are functionally tied to the central city, as defined in terms such as daily commuter traffic. Because large metropolitan areas typically consist of several groupings of counties meeting these criteria, the Census Bureau has defined a still higher-level aggregation, the Consolidated Metropolitan Statistical Area (CMSA), which is made up of two or more MSAs. When MSAs are included in a larger CMSA, they are denoted as PMSAs (Primary Metropolitan Statistical Areas).

Chicago's PMSA (population 8,272,768 in 2000) consists of the nine counties of northeastern Illinois. The boundaries extend well into the cornfields of central Illinois because MSAs are based on county lines rather than on population density. Chicago's PMSA is part of the Chicago-Gary-Kenosha CMSA (population 9,157,540 in 2000), which includes those two PMSAs plus the Kankakee PMSA. The CMSA thus includes thirteen counties and approximates the outer functional region focused on the city of Chicago. It includes large and small cities, villages, farms, and everything else in those counties.

THE TRADE MODEL

Chicago's region can be defined in many terms other than those used by the U.S. Census. What makes Chicago an important city has as much to do with its relations with the nation and the world as a whole as it does with its local area. Chicago's rise in the second half of the nineteenth century was based on the role it played in trade, commerce, and industry. As Carl Sandburg wrote in his memorable poem, "Chicago" (1914), the city was:

> Hog Butcher for the World,
> Tool Maker, Stacker of Wheat,
> Player with Railroads and the Nation's Freight Handler…

Very few hogs are butchered in the city these days, the tool-making business is on the decline, and, although Chicago's commodities traders are partly responsible for determining the price of wheat, not much actual grain is stacked in the city. Yet Chicago's trade role remains, and, in the era of globalization, it is more important than ever.

"Globalization" is a term often used negatively to describe a trend toward out-sourcing manufacturing activities to distant realms of cheap labor and lax environmental regulation. While these associations are fact in many cases, the products of globalized production are sold mostly in the same markets as before. Transportation systems have had to evolve to meet new challenges of hauling goods farther and faster. Because the products are valuable, and because suppliers and markets are more widely separated, there is also a need to expedite the flow of goods so they reach market when they are needed.

Chicago has lost manufacturing jobs through globalization, but it has gained in importance as a transportation center (Fig. 1.2). Shiploads of goods manufactured in Asian countries arrive at the ports of Los Angeles and Long Beach every day. Los Angeles is now the busiest seaport in the United States, largely because it is the first port of call for most ships plying the trans-Pacific lanes. Seattle or Vancouver often are the last ports of call for those same ships before they turn back across the ocean. Whether the business is exports or imports, railroads carry thousands of containers designed for truck, rail, or ship handling between Chicago and the West Coast ports every day as part of this global system.

New York and other East Coast ports conduct a similar business with Western Europe. European-made goods destined for Southern California pass through Chicago going west. Products originating in the western United States are funneled through Chicago bound for the East Coast or Europe. Atlanta, Jacksonville, and Miami are functionally tied to Chicago in a similar arrangement to the southeast. Globalization has helped place Chicago in the center of world trade as it never was before.

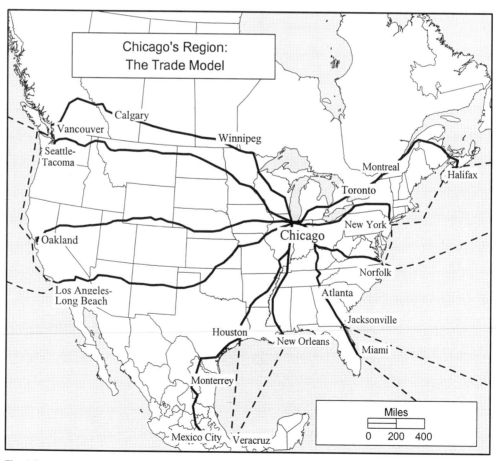

Fig. 1.2

The North American Free Trade Agreement (NAFTA) of 1994 has strengthened the north-south alignment of trade between Canada, the United States, and Mexico. Canada sells an ever-expanding array of products to Mexico just as Mexican factories produce more goods destined for sale in the United States and Canada. Chicago is central to much of this trade. Canadian companies have expanded their ownership in American railroads in recent years, not just to reach markets in the United States, but also to provide easier access to Mexican destinations. Chicago is the most important location where freight moving between Canada and Mexico is delivered by one railroad to another.

Chicago still is the "Nation's Freight Handler." It now has an expanded role handling the freight of the continent and the world as a whole. The city's relatively central location in the North American continent has something to do with its importance as a transportation center, although centrality alone does not account for all of Chicago's prominence. Many other cities have locations that are as central.

Chicago is the nation's railroad center today just as it has been ever since the 1850s. When railroad ventures were launched in the Eastern states during the mid-nineteenth century, railroad builders needed some western destination as a goal for their tracks to reach. Chicago, St. Louis, and various other cities fulfilled that role. Other railroad companies were founded by Chicagoans, largely for the purpose of building west into the prairies and plains to capture the trade of those places. The result was that most railroads either began or ended at Chicago. None of them merely ran through the city on their way east or west.

This circumstance made it necessary for nearly all railroad companies to interchange traffic at Chicago (Fig. 1.3). Railroad yards proliferated over the south and west sides of the city during the second half of the nineteenth century, and more were added in the twentieth. The scatter of railroad yards in the present era is the local-scale counterpart of Chicago as the point of North American intersection (Fig. 1.2). On magnification, the point becomes a tangle of tracks and yards, owned or operated by many separate companies, each having the purpose of originating traffic, terminating it, or interchanging traffic between lines.

Railroad companies entered a phase of corporate mergers during the 1960s. The trend continued until there are only about a half-dozen major railroads in North America at present. The pattern of mergers preserved the regional concentration of ownership, however, and with few exceptions railroads that operate east of Chicago do not operate to the west, and vice-versa. Many of the yards that these merged companies inherited from their predecessor companies are small and no longer fulfill the purpose for which they were intended.

It is an archaic arrangement, although one that might have continued for some time had Chicago not experienced a large increase in through-freight traffic begin-

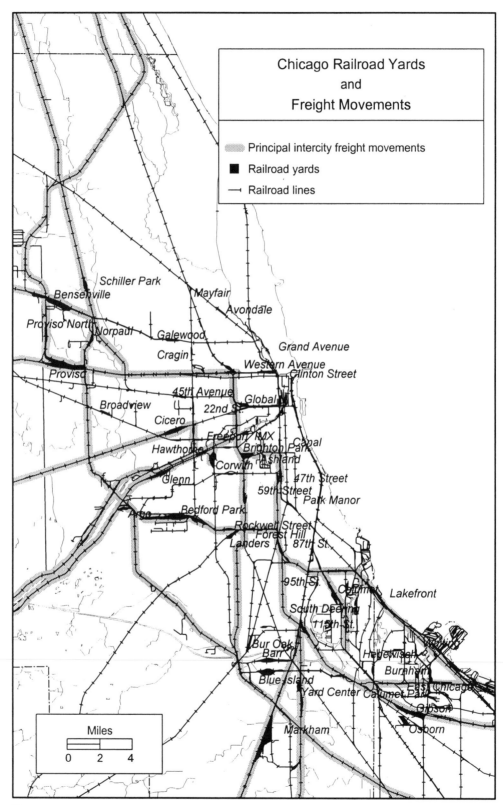

Chicago Railroad Yards and Freight Movements

Legend:
- Principal intercity freight movements
- Railroad yards
- Railroad lines

Schiller Park
Bensenville
Mayfair
Avondale
Proviso North
Norpaul
Galewood
Grand Avenue
Cragin
Western Avenue
Proviso
Clinton Street
45th Avenue
Global
Broadview
22nd St.
Cicero
Freeport IMX
Canal
Hawthorne
Brighton Park
Ashland
Corwith
Glenn
47th Street
59th Street
Park Manor
Argo
Bedford Park
Rockwell Street
Forest Hill
Landers
87th St.
95th St.
Calumet
Lakefront
South Deering
111th St.
Bur Oak
Barr
Hegewisch
Burnham
Blue Island
East Chicago
Yard Center
Calumet Park
Gibson
Markham
Osborn

Miles
0 2 4

Fig. 1.3

ning in the 1990s. Domestic parcel deliveries, automobiles moving from factory to showroom, coal needed to fuel electric power plants, frozen food, imports, exports, grain moving from farm to market, and a long list of other shipments must be accommodated in the system that is centered on Chicago.

Added to the bottleneck effect were the local demands for commuter rail service. Many of the same tracks needed for freight shipments are also commuter train routes. The mix of trains—some moving slowly, others rapidly—had the effect of clogging the whole system at morning and evening rush hours. In the late 1990s, Chicago railroads began to plan ways to make the interchange of traffic more efficient. Certain routes were selected for upgrading and improved connections (Fig. 1.3). It may take some years to remodel the entire system so that all of the uses demanded of it can be met more efficiently.

Because railroad yards consume a great deal of space they were built in what were peripheral locations at the time of their construction. The oldest yards are in the central city, the newest ones are located some distance out on the suburban fringe. Chicago's railroad yards have leapfrogged out past the suburban fringe in advance of high-density suburban growth, and they continue to exhibit that tendency. The latest additions to the pattern of yards are two massive terminals (not shown on Fig. 1.3) built in 2003 for intermodal truck-rail interchange traffic and for originating and terminating trains. The Burlington Northern-Santa Fe Railroad built Logistics Park south of Joliet, fifty miles south of downtown Chicago; Union Pacific constructed a new intermodal center, Global III, at Rochelle, seventy-five miles to the west. In time, these two widely separated sites will become focal points for much of the traffic shown in Fig. 1.2. In other words, they will become "Chicago" in terms of the trade model.

STUDYING CHICAGO

A book about Chicago and its region could be focused at many scales and undertaken from a variety of perspectives. Census metropolitan designations and the city's role as a trade and transportation center are just two of them. This book takes a topical and chronological approach to the subject of city and region because those two strategies have stood the test of time as useful ways of organizing what is known about places in a logical and accessible fashion.

The following two chapters, *The Chicago Model* and *Chicago's Neighborhoods*, present two themes of longstanding importance. Chicago has been a favorite model for social scientists who have studied cities systematically. The city's many neighborhoods are perhaps the folk counterpart of those models because they define urban space in the way Chicagoans themselves understand their community. Chapters 4-6 focus on physical geography. Although Chicago's site is one of the flattest of any large American city, minor variations in topography are closely related to prehistoric patterns of residence and livelihood. Activity surrounding the "Chicago portage" is intimately connected with the city's early history. Both presence of the lake and the effects of local topography influence patterns of residential desirability in the region.

Parts III, IV, and V (Chapters 7–15) are organized chronologically as a study in historical geography. The time scale becomes progressively finer through these chapters as the focus narrows down to developments that have the most relevance today. The three parts are organized roughly into time periods: Part III on developments through the end of the nineteenth century; Part IV on growth and change from the 1890s through World War II; and Part V on developments since that time, especially on race, ethnicity, and suburbanization.

Part VI focuses on three themes that emerge directly out of issues raised in the several preceding chapters: commuting to work, neighborhood change, and patterns of income distribution. Part VII (Chapter 19) concludes the textual component of the book by examining how Chicago changed from a geographically balanced geography that prevailed into the early twentieth century, to the skewed pattern of sectoral imbalances that now exists.

The more than 100 maps contained in this book illustrate many changes in Chicago's geography that have taken place over time. They are not merely illustrative, but rather are integral to understanding the arguments put forward in the text. Likewise, the photographs presented in *Chicago Portfolio: Where Geography and Photography Meet* following Part VII offer the reader a chance to experience Chicago's people and places through the eyes of some of the city's and nation's most accomplished photographer-artists, providing both artistic renderings of the city and visual context to the text.

In Chicago, as in all other cities and places, geography matters.

Chapter Two
THE CHICAGO MODEL

Chicago has been studied by more social scientists for a longer time than perhaps any other city in the world. Generalizations about "typical" city forms and functions often are traceable to statements first made about Chicago. Whether this circumstance arises because so many studies of Chicago have been made by people who published their findings, or whether there is some truly typical or average quality about the city that makes it a model for others, is a matter of debate.

What is "typical" about a city can be based on characteristics that the city has, but also on peculiar details that it lacks. The fewer unique factors that impinge on a city's form, the more easily it can appear generic and resemble others. Some brief comparisons between the Chicago, New York, and Los Angeles urbanized areas offer some perspective on the typicality issue.

NEW YORK CITY, LOS ANGELES, AND CHICAGO URBANIZED AREAS

In 2000, the New York City urbanized area had a population of nearly eighteen million and was spread out over 3,239 square miles (Fig. 2.1). Los Angeles, with a population of almost twelve million, was the nation's second largest and had an urbanized area of 2,012 square miles (Fig. 2.2). Chicago, third largest, had an urbanized area population of more than eight million and covered 1,624 square miles (Fig. 2.3). All three urbanized areas had population densities between 5,000 and 6,000 persons per square mile. Los Angeles, which is often identified with dispersion and sprawl, had the highest density of the three.

New York is rarely considered a typical city. Its urbanized area includes suburbs in three states. The five boroughs of its central city are separated by broad rivers that converge at the mouth of New York Bay. New York's prominence has much to do

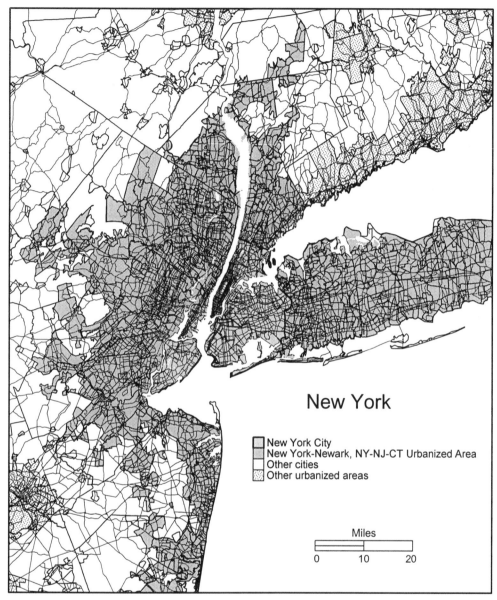

Fig. 2.1

with its excellent natural harbors, but the interruptions created by rivers and harbors limit access in various directions. Large areas of marsh and other swampy land cover what otherwise might be intensively occupied tracts on the New Jersey and Long Island fringes of the city. The uniqueness of New York's site overwhelms so many aspects of the city's geography that it could not be considered a model of how an average city might develop.

Los Angeles possesses more typical qualities than New York, but its site, too,

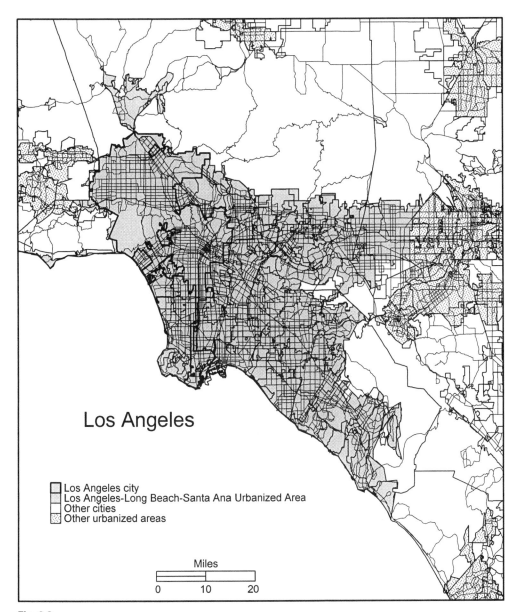

Fig. 2.2

exerts a powerful effect on the city's geography. Los Angeles is fringed by ranges of low mountains or hills on three sides. The Santa Monica, San Gabriel, Santa Ana, and San Bernardino mountains split the metropolitan area into zones of intense development separated by ridges of rugged ground that have much lower population densities. Although a network of freeways crosses the hills and mountains, topography spreads the city's population over a larger area than would be necessary if the land were flat.

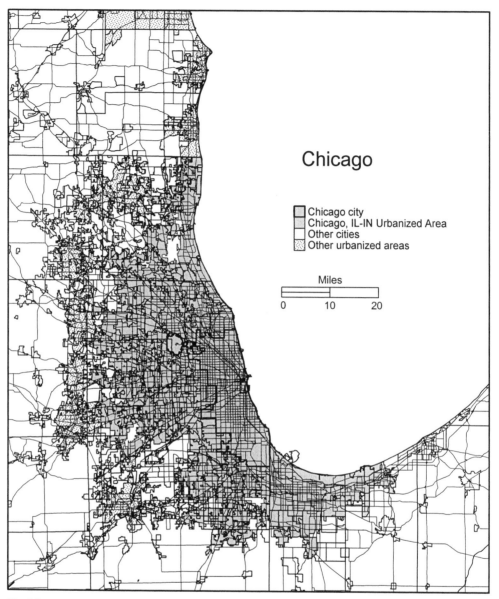

Fig. 2.3

Chicago's site is interrupted by neither rivers nor mountains. Built on the bed of a former glacial lake, Chicago expanded easily in all directions except, of course, over Lake Michigan. The land surface is flat, and there are no barriers to transportation. Highways and railroad lines radiate in all directions from the city except toward the northeast. The broad equalities and symmetries that Chicago's site allowed had a largely benign effect on the pattern of city growth. The roles played by economic and social factors are much easier to see in a city where nature offers few impediments to a uniform dispersion of activities. Chicago's physical environment definitely influenced the city's growth and expansion, but it neither interrupted nor limited it, as was true of New York and Los Angeles.

THE CHICAGO SCHOOL

Chicago's growth on a "flat and featureless plain" permitted economic and social factors to overwhelm natural site factors as influences on its geography. By the 1920s, the city's population was passing the three-million mark and growth was taking place in all directions. Many who came to Chicago possessed job skills that commanded high wages, but many others were relatively unskilled. The population of the United States was shifting from a rural to urban majority at that time. Migrants from rural areas and those coming directly from foreign countries were more likely to enter into one of the unskilled occupations. African-Americans, who had begun to move to Chicago in large numbers by the 1920s, lived in a narrow strip of densely settled blocks directly south of downtown.

The social dynamics of a growing city, with many new arrivals of various means, who came from widely different social-class backgrounds, presented a unique opportunity for scholarly research. Robert E. Park and Ernest W. Burgess, sociologists at the University of Chicago, pioneered the study of the "urban community" using Chicago as their laboratory. Along with Roderick D. McKenzie, the three men are those most identified with the Chicago School of Sociology, which focused on problems of urbanism and city life. Their concern with social problems drew special attention to inner-city neighborhoods where poverty, crime, and vice were common.

The inner city offered cheap housing of poor quality that was occupied by the most disadvantaged groups. The poor were forced to accept those conditions, but in time they might become more established and move out to newer areas of the city where housing was better and crime was less of a problem. Their places in the older, run-down parts of the city would then be occupied by new groups of immigrants who could not help but repeat the process. The city thus was seen as an evolving pattern of concentric rings representing the stages of outward movement, social evolution, and economic betterment over time.

The Park-Burgess-McKenzie model had a powerful influence on all subsequent theories about the nature of cities (Fig. 2.4; the present-day urban street grid has been superimposed on their model). The concentric ring model was both a description of Chicago as it existed and a theory about urban social processes. Labels such as "Bright light area," "Little Sicily," "Deutschland," "Chinatown," "Underworld," "Black Belt," and "'Two Flat' Area" were harsh but perhaps fitting for a geography of Chicago in the era of Al Capone. "Residential Hotels" and "Roomers" represented areas of the city where recent arrivals, especially younger unmarried individuals,

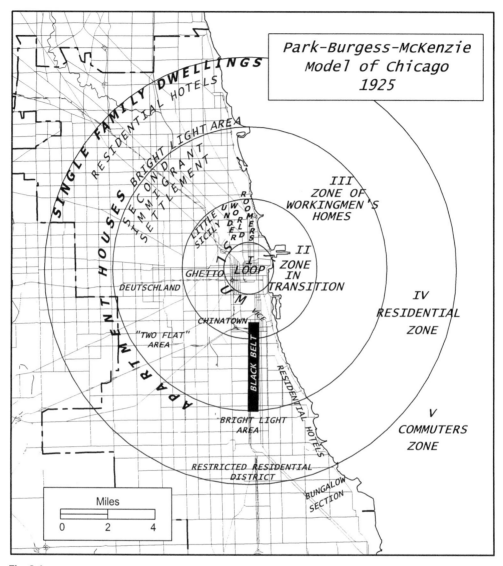

Fig. 2.4

were found. The "Bungalow Section" was in a more distant zone where single-family dwellings were common. Residential segregation was incorporated into the model in the designation of a "Restricted Residential District."

The Park-Burgess-McKenzie model focused on social class, which did have a concentric zonation somewhat like they described, but their model largely ignored economic variables such as accessibility and the price of land. Homer Hoyt, an economist and urban planner writing in the 1930s, was more interested in the linear patterns of high-priced housing in Chicago. The city's lakefront was a continuous zone of expensive housing that radiated north of downtown. Similar high-rent areas

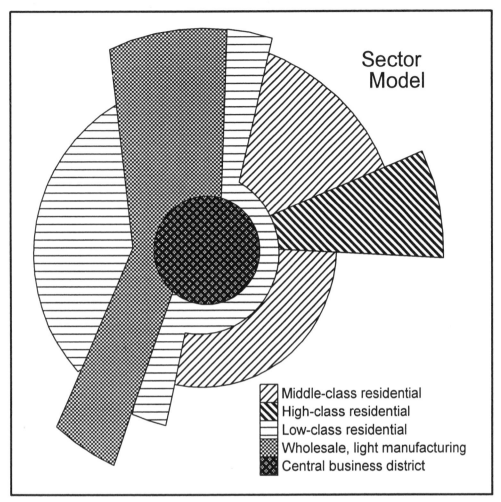

Fig. 2.5

(which were to disappear in later years) extended directly west and south of the city's center. Although Hoyt's model was narrowly focused, it suggested the importance of radial (sectoral) rather than concentric growth within the city.

In 1945, geographers Chauncy D. Harris and Edward L. Ullman published their much-cited study, "The Nature of Cities." They described the concentric zone model in roughly the same terms that Park, Burgess, and McKenzie had used, but they reworked Hoyt's model into a general scheme of urban land use based on accessibility (Fig. 2.5). The sector model, as it became known, is credited to Hoyt, although it was Harris and Ullman who elevated it to a general description of the city. The diagram they produced was based more on the example of Salt Lake City than on Chicago, but the central idea of the model could apply to any city with a capitalist land market.

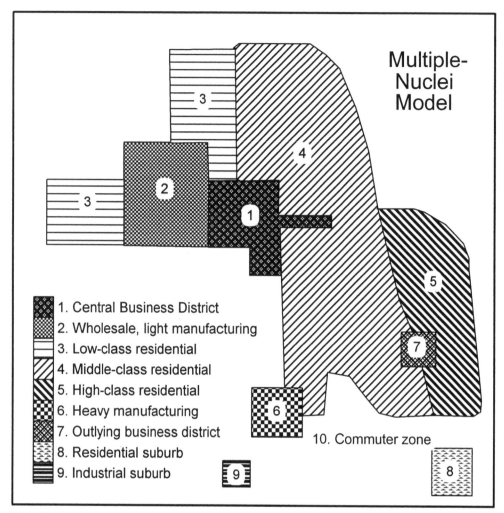

Multiple-Nuclei Model

1. Central Business District
2. Wholesale, light manufacturing
3. Low-class residential
4. Middle-class residential
5. High-class residential
6. Heavy manufacturing
7. Outlying business district
8. Residential suburb
9. Industrial suburb
10. Commuter zone

Fig. 2.6

The basis of sectoral (rather than concentric) zonation was economic. Land uses competed for high-priced land. Activities such as wholesaling and light manufacturing needed transportation access, which meant they would follow a linear pattern along transportation corridors. Low-class residential areas would border wholesaling and manufacturing. High-class residential areas were sectoral but also peripheral to the city center. Like the Park-Burgess-McKenzie model, the sectoral model was predictive, but it was formulated in terms of land prices rather than social mobility. How much land each activity needed and how much each could afford to pay for centrality determined where the activity would locate.

In the same paper, Harris and Ullman described a third idealized city form, the multiple-nuclei model (Fig. 2.6). It combined features of both concentric zonation and radiating sectors, but emphasized that large cities were influenced by unique factors that changed their form in unpredictable ways. Although the multiple-nuclei model did not look like Chicago, it was inspired, at least in part, by Chicago examples. Hyde Park, on Chicago's South Side, illustrated category 7, the outlying specialized-function district. The Clearing Industrial District on the Southwest Side illustrated category 9, the industrial suburb. High-priced amenity residential areas along Chicago's lakefront were incorporated as category 5, high-class residential. Heavy manufacturing, category 6, was exemplified by Chicago's stock yards and meat-packing complex.

The multiple-nuclei model offered a more satisfactory description of urban land use, but it had no power to predict where the various nuclei might be found. The model was so general it might be applied to any city, although the separate nuclei would have to be identified prior to applying the model. It did not deny the importance of the other models, but rather suggested that site-specific factors must be incorporated. The three models have endured as the most commonly used hypothetical examples of city form, each emphasizing a different aspect of urban structure.

CHICAGO II

Although the influence of the Chicago School eventually declined, the scholarly study of cities continued, especially under the leadership of geographer Brian J. L. Berry at the University of Chicago in the 1960s. This second Chicago school, which has been succinctly labeled "Chicago II," took a more mathematical approach to the subject (Shearmur and Charron, 2004). Berry and his students used the ever-growing array of statistical data collected by the census and analyzed it using multivariate statistical techniques.

Following the work of quantitatively oriented urban sociologists, they identified clusters of variables ("factors") that were thought to represent independent dimensions of social structure. Age, family size, and income typically were found to vary systematically over the city. Social variables, including occupation, class, and status, varied in radiating sectors. Income, family composition, and age varied more in concentric fashion.

Population density was studied perhaps more than any other single indicator. In the 1940s, it had been discovered in Europe that the rate of decline in population density away from a city's center varied according to a city's size and shape (Clark, 1951). The rate of change in density with respect to distance from the city center was proportional to density; that is, the higher the density, the steeper the curve. The model took the form

$$p(x) = p(0)\exp(-\lambda x)$$

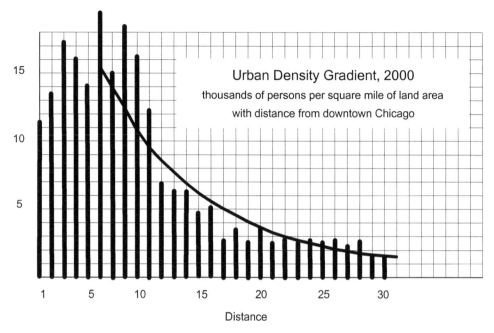

Fig. 2.7

where p(*x*) is population density *x* miles from the city center, p(o) is central density, and λ is the density gradient. This simple model allowed comparison of density gradients and central densities between cities and it was eventually applied to cities worldwide.

The urban density model was interpreted primarily in economic terms. Central locations are the most desirable because they offer the greatest access to all parts of the city. Those who need centrality bid up the price of land, which reaches a peak value near the city center. Because central locations are the most expensive, those who need more land and/or have less ability to pay accept locations that are more remote from the core. Commercial land uses have the greatest need for central location and hence outbid all other uses, including residential, for the most accessible sites. A population density "crater," where residential densities were lower, was found to exist near the city center. The model provided a good approximation to residential densities in Chicago and many other cities.

Density curves for Chicago based on 1960 data were summarized by Berry and Horton (1970). Population density gradients ranged from λ = -.03 on the North Side of the city to λ = -.08 on the South Side. Estimated average density at a distance of six miles from the city center was p(6) = 25,322 per square mile.

For purposes of comparison, the urban density model was recalculated for 2000 using census-block group data (Fig. 2.7). Populations were aggregated by one-mile concentric zones out to a distance of thirty miles. Water areas were excluded from density computations. The overall (nondirectional) value of the density gradient was found to be λ = -.096. Estimated density six miles from the city center was $p(6)$ = 15,410 persons per square mile. Comparison of the two sets of values suggests that Chicago declined in central density and had a steeper gradient in 2000 compared with 1960. Both changes would be expected in an urban area that is either shrinking in size or growing slowly.

Chicago's 2000 density map demonstrates the relevance of both concentric and sectoral patterns within the city (Fig. 2.8). Areas within six miles of the city's center have lower population densities. Residential space is crowded out where highways and railroads converge because of the large amount of land needed for transportation routes. Residential densities are at a maximum about six miles from the city center in a zone of high-rise or at least multistory apartment buildings. Low-density sectors along expressways, railways, and rivers extend beyond the six-mile radius. Apart from the low-density corridors, Chicago's population is evenly distributed around the city in the six- to ten-mile zone.

A significant reduction in density takes place roughly ten miles from downtown. The drop coincides with a shift away from multistory apartment buildings to two-flats and single-family homes on small city lots. A second drop takes place about sixteen miles out where there is a transition to larger residential properties. Chicago's population density gradient is difficult to define beyond a radius of thirty miles. Satellite cities at that distance (Waukegan, Elgin, Aurora, and Joliet) have their own population peaks and density gradients.

The urban-density model, like the concentric zone, sector, and multiple-nuclei models that preceded it, was an attempt to summarize the complexities of urban geography in a manner that allowed direct comparisons between cities. That the example of Chicago was so influential in these models probably is due mostly to the fact that they were the work of Chicago-based scholars. In recent years, others have argued that Los Angeles deserves a more central role in this respect (Dear, 2002; Scott and Soja, 1996).

The fact that a given generalization applies unequally to the cases it purports to explain is not surprising. Focusing on Chicago alone, it seems fair to conclude that the models proposed over the past seven or eight decades have had, and continue to have, some validity. They are useful and illustrative, but they are neither "laws" of urban form nor explanations for even a large share of what is geographically interesting about the city. Chicagoans do not view their city in terms of models such as these, although they do have a distinctive and highly geographical way of identifying and differentiating the urban space they occupy. For Chicagoans, it is "the neighborhoods" that matter, and no human geography of the city can be meaningful without them.

Population
Density,
2000

Persons per square mile
Fewer than 1000
1000 to 5000
5000 to 10,000
More than 10,000

10 miles

5 miles

Miles
0   2   4

Fig. 2.8

Chapter 3
CHICAGO'S NEIGHBORHOODS

CHICAGOANS BELIEVE IN NEIGHBORHOODS. People identify their place of residence in neighborhood terms, real estate agents rely on neighborhood names when advertising properties for sale, and the City of Chicago uses neighborhood boundaries to organize various administrative functions.

The neighborhood idea has both positive and negative uses. As a source of place identity, neighborhoods generally have positive qualities associated with inclusiveness and belonging. There also can be negative effects, as when the focus is on boundaries or lines of demarcation and when exclusion is the purpose of invoking a neighborhood's identity. Neighborhoods are regarded as having near-permanent status. Rarely are they relabeled or redrawn, even though they undergo nearly constant physical and social change over time.

NEIGHBORHOOD DESIGNATION

The idea of neighborhoods in Chicago goes back to the Chicago School sociologists of the 1920s. The university's Social Science Research Committee (following the work of its predecessor, the Local Community Research Committee) named and defined the boundaries of seventy-five community areas of the city. The set of community areas covered the entire city in a non-overlapping fashion. Five criteria were used to identify and demarcate their boundaries: (1) settlement, growth, and history of the area; (2) local identification with the area; (3) local trade area; (4) distribution of membership of local institutions; and (5) natural and artificial barriers (Local Community Fact Book, 1980, xix).

The five criteria for defining Chicago's community areas were similar to those used by rural sociologists who were delimiting rural trade areas and communities of farmers during the 1920s and 1930s. In the city, as in the countryside, the notion of community as an areal grouping within a larger settled territory was considered to be innate. Discovery of the community areas—even their boundaries—could be accomplished by local investigation supplemented with the analysis of census data. Neighborhoods of this sort never were to be imposed by outsiders, let alone by government action; rather, they were already there, waiting to be discovered by trained researchers for the purpose of understanding local social organization.

Although neighborhoods and community areas so defined were informal entities, they were adopted later by two levels of government and thus acquired official status. When the U.S. Bureau of the Census undertook the task of subdividing cities into census tracts, Chicago's community areas became part of the process. Chicago's census tracts are subdivisions of its seventy-five community areas; that is, census tracts aggregate upward into community areas and never are split between two or more such areas. This was a major improvement over the Census Bureau's practice through 1920, which was to report statistics for urban areas according to city wards. Ward boundaries change frequently due to reapportionment. Data published by wards were not useful for making comparisons between censuses.

The City of Chicago also adopted the seventy-five community areas as its official neighborhoods for city planning purposes. In the 1960s, two more community areas were recognized, bringing the total to seventy-seven, but no changes have been made since that time. The City of Chicago officially recognizes two levels of subdivision. The seventy-seven community areas are supplemented with a finer-scale division, "Chicago neighborhoods," of which there are 172. Informal labels and names abound, however, and the list of 172 could well be much longer. Neighborhood names in common usage today include some that are not officially recognized by the City of Chicago (Fig. 3.1; the map omits a number of small official neighborhoods that would be difficult to show at this scale).

The term "neighborhood" enjoys common and frequent usage in Chicago. It is an inclusive category that embraces community areas, official neighborhoods, and numerous informal designations of territory. Neighborhoods range in size from the land surrounding a street-corner park, to tracts of twenty square miles or more. The tendency to recognize neighborhoods is strongest in residential areas and weakest in zones where industrial land use predominates. Many of the 172 official neighborhoods can be aggregated upward into the seventy-seven community areas, although others are split between two community areas, making their aggregation into community areas impossible.

Community areas are the obvious entities to use in conjunction with census data. The real test is whether they are useful for portraying socioeconomic patterns in the city (they are), and they will be used in the balance of this chapter to give a summary

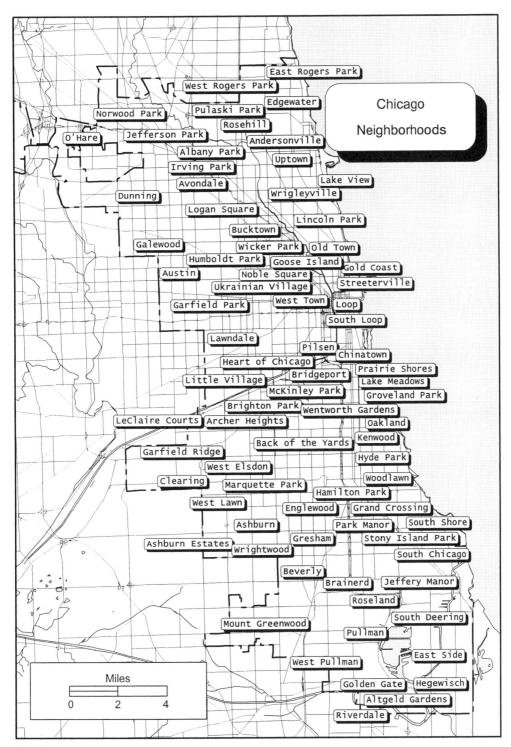

Chicago Neighborhoods

East Rogers Park
West Rogers Park
Edgewater
Norwood Park
Pulaski Park
Rosehill
Jefferson Park
Andersonville
O'Hare
Albany Park
Uptown
Irving Park
Lake View
Avondale
Wrigleyville
Dunning
Logan Square
Lincoln Park
Bucktown
Galewood
Wicker Park
Old Town
Humboldt Park
Goose Island
Gold Coast
Austin
Noble Square
Ukrainian Village
Streeterville
Garfield Park
West Town
Loop
South Loop
Lawndale
Pilsen
Chinatown
Heart of Chicago
Prairie Shores
Little Village
Bridgeport
Lake Meadows
McKinley Park
Groveland Park
Brighton Park
Wentworth Gardens
LeClaire Courts
Archer Heights
Oakland
Back of the Yards
Kenwood
Garfield Ridge
Hyde Park
West Elsdon
Woodlawn
Clearing
Marquette Park
Hamilton Park
West Lawn
Englewood
Grand Crossing
Ashburn
Park Manor
South Shore
Ashburn Estates
Gresham
Stony Island Park
Wrightwood
South Chicago
Beverly
Brainerd
Jeffery Manor
Roseland
Mount Greenwood
South Deering
Pullman
West Pullman
East Side
Golden Gate
Hegewisch
Altgeld Gardens
Riverdale

Miles
0   2   4

Fig. 3.1

description of Chicago's current human geography. Just four socioeconomic variables are needed to produce a regionalization: the percentages of the population classified as white-collar, African-American, foreign-born, or blue-collar. The four-fold grouping is neither mutually exclusive nor exhaustive, and hence it does not form a well-defined classification; but each of the four has a pattern that highlights a wide variety of other factors underlying neighborhood differentiation.

WHITE-COLLAR NEIGHBORHOODS

White-collar neighborhoods can be identified in a map of professional and managerial workers as a percentage of the employed population (Fig. 3.2). Until fairly recent times, the populations of these neighborhoods were largely of Euro-American heritage, although people of Asian background now constitute a growing fraction of the population. Forest Glen and Beverly, at the extreme northwestern and southwestern corners of the city, respectively, are long-established white-collar neighborhoods that house many doctors, lawyers, and other professionals. Employment in colleges and universities is associated with high white-collar percentages in North Park (North Park College and Northeastern Illinois University) and Hyde Park-Kenwood (University of Chicago).

The largest aggregation of white-collar professionals is the Chicago lakefront. Portions of the Uptown neighborhood, Lake View, Lincoln Park, and the Near North Side have especially high concentrations. These are high population-density neighborhoods where many people reside in multistory apartment and condominium buildings. The pattern of high status and high incomes continues south beyond the Loop where new apartment buildings or converted commercial structures make up a housing stock that is expensive to buy or rent.

The Lake Michigan shoreline is regarded as Chicago's most attractive landscape feature. While the lake itself is fringed by a strip of parks and beaches on public land, high-rise apartments set back from this zone have an unobstructed view of the lake. High status and high incomes have been associated with these neighborhoods for much of the city's history, even though the housing stock has changed several times.

Newer concentrations of white collar employment have appeared well back from the lakefront. The North Center and West Town areas are two where white populations have grown substantially in the past two decades. The Near West Side white-collar neighborhood includes the Rush-Presbyterian-St. Luke's medical complex and the University of Illinois at Chicago.

Lakefront land is valuable, and it is likely to be even more associated with white-collar/professional employment in years to come. Both the Douglas and Oakland neighborhoods are split between lower-income tracts that are predominantly poor

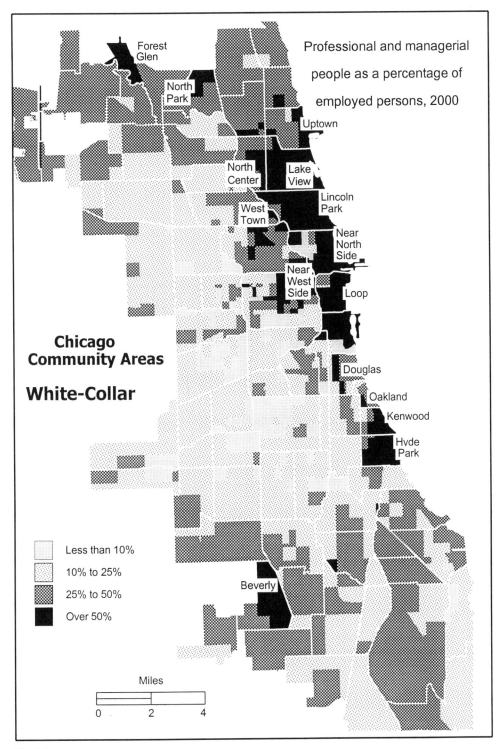

Professional and managerial people as a percentage of employed persons, 2000

Forest Glen

North Park

Uptown

North Center

Lake View

Lincoln Park

West Town

Near North Side

Near West Side

Loop

Chicago Community Areas

White-Collar

Douglas

Oakland

Kenwood

Hyde Park

Less than 10%

10% to 25%

25% to 50%

Over 50%

Beverly

Miles

0   2   4

Fig. 3.2

and African-American, in contrast with clusters of new housing that are attractive to high-income households.

AFRICAN-AMERICAN NEIGHBORHOODS

More than half of Chicago's neighborhoods have a large African-American population (Fig. 3.3). Apart from some census tracts with a diverse ethnic mixture on the city's Far North Side, where African-Americans increased in numbers during the 1990s, most areas of Chicago that have sizable African-American populations are more than ninety percent black. To speak of "concentrations" of African-Americans substantially understates the extent to which whites and blacks live in totally separate parts of the city.

Chicago's oldest African-American neighborhoods grew immediately south of the downtown area. This was the "black belt" identified by Park, Burgess, and McKenzie in the 1920s (Fig. 2.4). The Douglas, Oakland, Grand Boulevard, Fuller Park, and Washington Park neighborhoods had majority African-American populations by 1930, and most were already in the over-ninety-percent category. Large black populations appeared in the Near West Side neighborhoods at the same time. Although Chicago's South Side black population has always been larger and covered more area, African-American majorities appeared on the West Side of the city only slightly later. During the 1950s, blacks moved into North Lawndale and the two Garfield Park neighborhoods. Humboldt Park and Austin followed the trend during the 1960s.

The 1950s saw roughly a doubling of the area occupied by African-Americans on the South Side, which included expansion into the Englewood, Greater Grand Crossing, Washington Heights, Hyde Park, and Chatham neighborhoods. All but Hyde Park, which retained a mixed population, became more than ninety percent black within a decade. During the 1960s, South Shore, South Chicago, Calumet Heights, Avalon Park, Pullman, West Pullman, Auburn-Gresham, and West Englewood had population turnovers of ninety percent or more in the shift from white to black.

A few areas that have a substantial African-American presence were not the product of growth and expansion of adjacent black populations. Public housing projects built in the Near North Side (Cabrini Green), Garfield Ridge (LeClaire Courts), and Riverdale (Altgeld Gardens) neighborhoods dispersed the black population but did not lead to subsequent local expansion in later years. Morgan Park, on the Southwest Side, had a small African-American population by 1920, which grew in size during later decades. Apart from these few exceptions, nearly all of the neighborhoods classified in the African-American category in 2000 were the product of black immigration and white abandonment during a single, continuous upheaval that cast a shadow over the city for more than six decades.

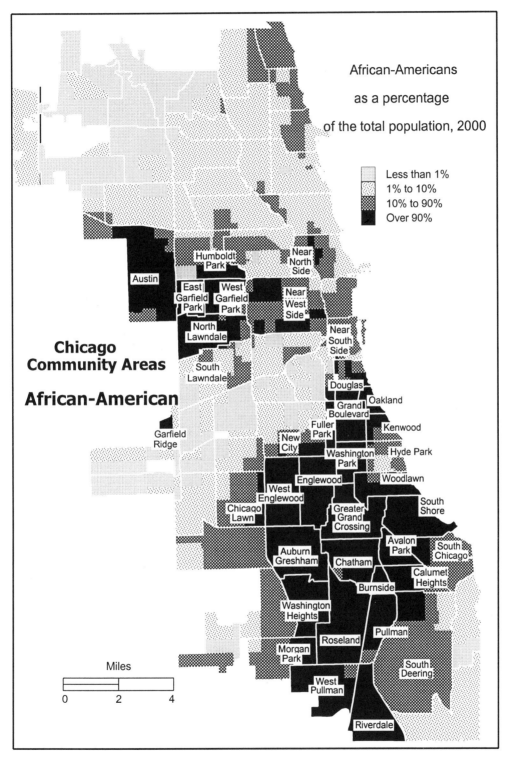

African-Americans

as a percentage

of the total population, 2000

Less than 1%
1% to 10%
10% to 90%
Over 90%

Humboldt Park

Near North Side

Austin

East Garfield Park

West Garfield Park

Near West Side

North Lawndale

Near South Side

Chicago Community Areas

South Lawndale

Douglas

African-American

Grand Boulevard

Oakland

Fuller Park

Kenwood

Garfield Ridge

New City

Washington Park

Hyde Park

Englewood

Woodlawn

West Englewood

South Shore

Chicago Lawn

Greater Grand Crossing

Auburn Greshham

Avalon Park

South Chicago

Chatham

Calumet Heights

Burnside

Washington Heights

Pullman

Roseland

Morgan Park

South Deering

West Pullman

Riverdale

Miles

0   2   4

Fig. 3.3

Chicago has long been a city of immigrants, although the countries that have contributed the largest numbers to the city's population have changed substantially over time (Fig. 3.4). People born in western and northern Europe dominated until the 1890s, when large numbers from eastern and southern Europe began arriving. These were years when Chicago was expanding greatly in area as well as in population, and to some extent new immigrants took up residence in areas that had not been settled in earlier times. Most European groups, however, followed a well-established pattern of early residence in the inner city followed by outward expansion and eventual movement to the suburbs.

By 2000, Europeans accounted for less than one-fourth of Chicago's foreign-born population. Nearly half the European immigrants were Polish. Poles established an early pattern of ethnic concentration in sectors radiating west from downtown Chicago. Generation by generation, they continued to expand over more area as the population grew from immigration. Once established, individuals and families moved to newer neighborhoods on the Northwest or Southwest sides of the city.

Today more than half of all Polish immigrants in the Chicago area were living outside the central city in dozens of suburbs. The O'Hare, Dunning, Montclare, Belmont-Cragin, Portage Park, Hermosa, and Logan Square neighborhoods were the remaining areas of large a Polish population on Chicago's Northwest Side. Archer Heights, West Elsdon, West Lawn, Garfield Ridge, and Clearing are the counterparts of those neighborhoods on the Southwest Side.

The greatest change in Chicago's foreign-born population was the increase in immigration from Mexico that began in the 1950s. Mexicans now account for forty-six percent of Chicago's foreign-born. Other Latin American sources contribute an additional ten percent. Of the three early Hispanic areas (Southeast, Southwest, Northwest), the Northwest Side concentration has expanded most rapidly since 1970. West Town, Avondale, Logan Square, Irving Park, Albany Park, and Hermosa grew from both Mexican and Puerto Rican immigration. The Lower West Side, McKinley Park, New City, South Lawndale, Brighton Park, Archer Heights, and Gage Park neighborhoods experienced a substantial Hispanic (largely Mexican) influx at the same time. Mexican populations in Chicago's East Side neighborhood, on the far Southeast Side of the city, are part of a concentration of the Mexican-born that extends eastward into Indiana. Many of the neighborhoods with large Hispanic populations today were Polish one or two generations ago.

A third type of ethnic concentration is found in the Far North Side neighborhoods, where immigrants from Asia are most concentrated. Asian populations in the West Ridge, Lincoln Square, North Park, and Albany Park neighborhoods have grown steadily in size, and they have expanded eastward into Rogers Park, Edgewater, and

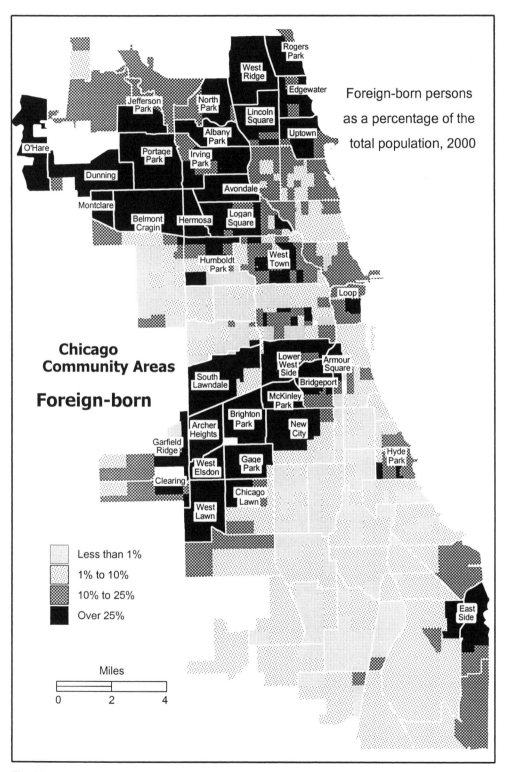

Foreign-born persons as a percentage of the total population, 2000

Chicago Community Areas

Foreign-born

Less than 1%
1% to 10%
10% to 25%
Over 25%

Miles
0   2   4

Fig. 3.4

Uptown. Immigrants from India, Korea, and the Republic of the Philippines are represented in large numbers, but nearly two dozen other Asian countries have a substantial presence in these areas as well. The old Chinatown neighborhood on the Near South Side accounts for the large foreign-born population in the Armour Square neighborhood. People from China account for the largest share of Hyde Park's foreign-born component.

Asian immigrants have an established presence in all of these areas, but their pattern within the city of Chicago is of lesser importance than their recent habit of skipping the central city and taking up residence in the suburbs immediately following immigration. Today, immigrants from India, Korea, and the Philippines are more than twice as likely to live in one of the many suburbs in Cook County as they are to live in the city of Chicago.

BLUE-COLLAR NEIGHBORHOODS

Chicago thrived for decades on its economic base of manufacturing industries. Added to the blue-collar jobs in manufacturing were many others in transportation and construction that provided employment for generations of newcomers and established residents alike. The map of blue-collar occupations in 2000, defined as the percentage of employed workers engaged in manufacturing, construction, or transportation, reveals a fourth perspective on Chicago's geography (Fig. 3.5). While not all occupations are readily definable as being either white-collar or blue-collar, it is not surprising that blue-collar workers are largely absent in the areas where the white-collar professional and managerial groups dominate (Fig. 3.2).

Nor is the blue-collar pattern independent of other concentrations just reviewed. Today, as in the past, many blue-collar neighborhoods are also those where the foreign-born are most heavily represented. Hispanic populations, broadly defined, and immigrants from Mexico, in particular, are often employed in manufacturing and construction jobs. Some African-American neighborhoods fall in the blue-collar category as defined by employment in these three sectors, but others do not.

What is most evident is the contrasting pattern between white-collar and blue-collar as a cleavage between neighborhoods near the lakefront and those that are remote from it. Most of Chicago's manufacturing industries that were put in place from the late nineteenth century through the middle of the twentieth grew around the western and southern margins of the city, as the sector model predicts. Apart from the Calumet Harbor industrial area on the Southeast Side (represented in Fig. 5.5 by the South Chicago, East Side, and Hegewisch neighborhoods), employment in blue-collar occupations was uncommon near the lakefront, which has remained primarily residential ever since the 1870s.

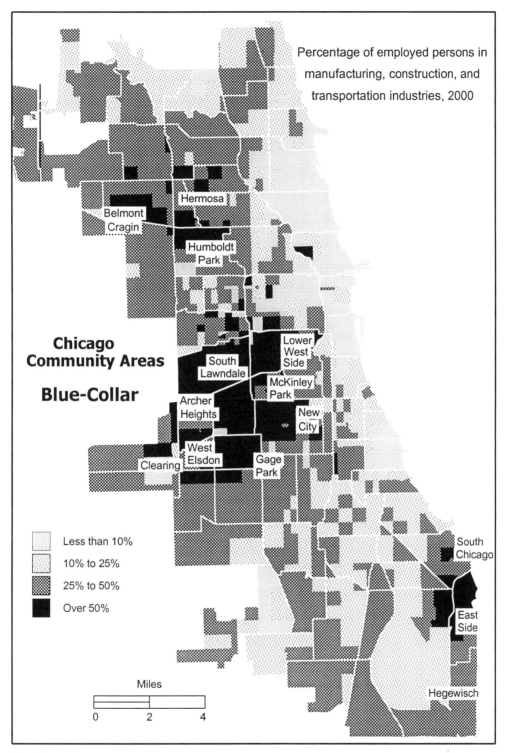

Percentage of employed persons in manufacturing, construction, and transportation industries, 2000

Chicago Community Areas

Blue-Collar

Hermosa

Belmont Cragin

Humboldt Park

Lower West Side

South Lawndale

McKinley Park

Archer Heights

New City

West Elsdon

Gage Park

Clearing

South Chicago

East Side

Hegewisch

Less than 10%

10% to 25%

25% to 50%

Over 50%

Miles

0   2   4

Fig. 3.5

Most of Chicago's neighborhoods thus can be identified in terms of high percentages in the white-collar, blue-collar, foreign-born, or African-American categories. Only a few scattered neighborhoods at the city's western extremities do not appear on any of the four neighborhood maps. They are a mixture of white-collar and blue-collar, in terms of occupation, and typically have large numbers of foreign-born. Their more varied compositions more closely reflect the suburbs to which they are adjacent than any patterns characterizing the central city.

The four variables used to construct the neighborhood maps reflect many other indicators that might be used to differentiate social areas within the city, such as income, education, age, housing quality, and home ownership. Chicago is by no means unique in the fact that its internal geography can be described in terms of a limited number of statistical variables. These few attributes by no means constitute a total description of the city's character, however, let alone its evolution over time as an urban place.

The next three chapters focus on patterns of climate, landforms, and surface drainage in the Chicago region. The physical environment often is treated as a constant set of background conditions that influence human settlements. While this view is often correct, Chicago's physical environment has not remained constant since the first people arrived. Environmental changes independent of human influence have mingled with the changes that people have brought about in their surroundings. Even at a human time scale, nature has not been a constant.

Part II

PHYSICAL GEOGRAPHY

Chapter 4
CHICAGO'S CLIMATE

CHICAGO'S CLIMATE IS THE HUMID, continental variety with warm summers, cold winters, and a fairly even distribution of precipitation throughout the year. In terms of analogs elsewhere in the world, Chicago's climate resembles that of Bulgaria or Romania, at the same latitude on the Black Sea. Typical of continental climates, Chicago's weather is variable, and individual years can be much warmer, wetter, cooler, or drier than average. Because the air masses producing seasonal and daily variations in the weather originate in distant source areas, such as northern Canada, the Gulf of Mexico, or the Pacific Ocean, short-term departures from seasonal averages are common.

LAKE EFFECT

"Lake effect" is an important secondary influence on Chicago's climate. The city's location on Lake Michigan subjects it to three influences that diminish with distance from the lake. At the seasonal level, Lake Michigan has a local warming effect in winter and a cooling effect in summer. On a daily basis, proximity to the lake can mean warmer overnight low temperatures but cooler daytime highs. Because most midlatitude weather systems move from west to east, Chicago feels these lake effects somewhat less than do communities in Indiana and Michigan on the east side of Lake Michigan.

The main explanation of lake effect comes from the physical concept of "specific heat," which is the amount of energy needed to produce a given temperature change in a substance. Fresh water, as in Lake Michigan, has a very high specific heat, meaning that it takes a large amount of energy to produce a given change in its temperature. Soil, rock, pavement, and other components of the land surface have much lower specific heat values; hence, a given energy input produces a greater change in their temperatures. All land and water surfaces at a given location are subject to the same energy inputs from the sun. It takes a longer time for a given amount of incoming solar energy to raise the temperature of water than to raise the temperature of land. The same process operates in reverse when energy inputs are decreased, and thus water loses its heats less rapidly than land.

The result is that land temperatures increase and decrease, on a daily as well as seasonal basis, more rapidly than water temperatures. Land areas very close to a body of water experience temperature changes more like those of the water than of the land, with the same lag effect. Chicago's lakefront can experience cooling daytime breezes on a hot summer day when winds are blowing off the lake, but also warmer overnight temperatures because the lake gives up its heat to the atmosphere less rapidly. Lake effect is illustrated in the map of January daily minimum temperatures around the Chicago area (Fig. 4.1). These lowest average temperatures of a typical year are increased roughly 4° F, from a chilly 12° F to a slightly less extreme 16° F, through the lake effect.

THE URBAN HEAT ISLAND

The map also illustrates another secondary effect, that of the "urban heat island." Cities typically are warmer than the rural areas around them because of factors such as air pollution, heat loss from buildings, and exhaust emissions that warm the urban environment. The added warmth produced by the lake effect is further enhanced by the heat-island effect in downtown Chicago. In northwest Indiana, where there is little or no urban heat island, the warming caused by Lake Michigan does not extend as far inland.

Effects of the urban heat island are well illustrated in the map of cooling-degree days (Fig. 4.2). Every degree that the average temperature for a given day exceeds 65°F is defined as one cooling-degree day. Summed over the warm season of the year, the total number of cooling-degree days is a measure of the demand for air conditioning. The total value is less than 600 near Lake Michigan outside the city. But away from the lake and toward the city, values can be nearly double that amount. The greatest demand for air conditioning is on the Southwest Side of Chicago, where the urban heat island effect is largest and where cooling lake breezes are rarely felt. Lakefront areas north of Waukegan and in extreme southwestern Michigan have the lowest demand for air conditioning in the greater Chicago region.

CLIMATE AND TOPOGRAPHY

In addition to lake effect and the influence of the urban heat island, elevation and topography can exert an influence on the local climate (Fig. 4.3). Annual precipita-

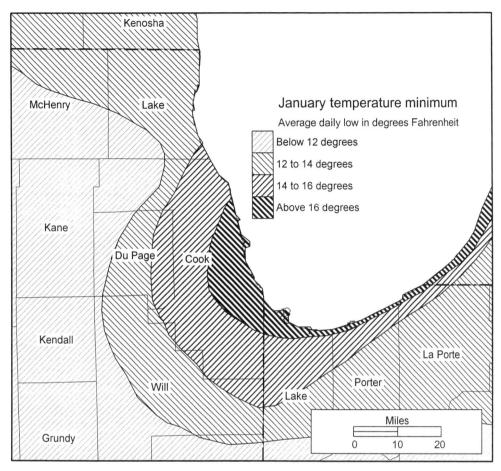

Fig. 4.1

tion in the Chicago area averages around thirty-seven inches (38.35″ at Midway Airport and 36.27″ at O'Hare Airport). Precipitation declines generally northward from Chicago, even along the lake. Cooler air holds less moisture. The west shore of Lake Michigan north of the city, upwind from the lake effect, is the driest portion of the metropolitan area.

Topography interacts with weather patterns because air moving upslope is cooled as it rises. Even a small increase in elevation can increase the chance of precipitation from a given air mass. This is illustrated south and west of Lake Michigan, where bordering glacial moraines produce enough lifting that additional precipitation results. Precipitation along the crest of the Valparaiso Moraine in Indiana averages one or two inches per year more than nearby areas to the north which, although they are on Lake Michigan, lie at a lower elevation. Northwesterly winds pick up moisture-laden air over the lake, which is then released as the air moves upslope.

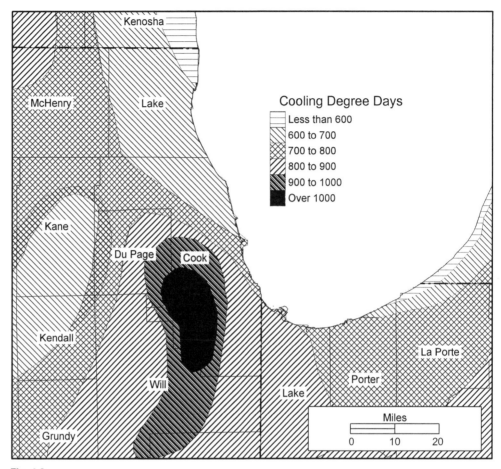

Fig. 4.2

Snowfall totals vary widely around the Chicago area (Fig. 4.4). Proximity to Lake Michigan generally means more snow, although wind direction substantially modifies the lake's effect. Northerly winds draw moisture over Lake Michigan and typically produce heavier snowfalls on the South Side of Chicago than on the North Side. But north-south differences are small compared with contrasts between eastern and western shores of the lake. At the lake's edge, southeastern Michigan receives more than twice as much snow as northern Cook County. Temperatures over the lake often are warmer than land temperatures in winter, which can produce added evaporation of moisture into the atmosphere. The "snowbelt" begins around Evanston, where land jutting into Lake Michigan intercepts southward-moving air masses that have absorbed moisture over the lake. The effect increases counterclockwise (eastward) around the southern end of the lake. In Michigan, winds from every direction except the southeast pick up moisture from the lake and deposit it as snow.

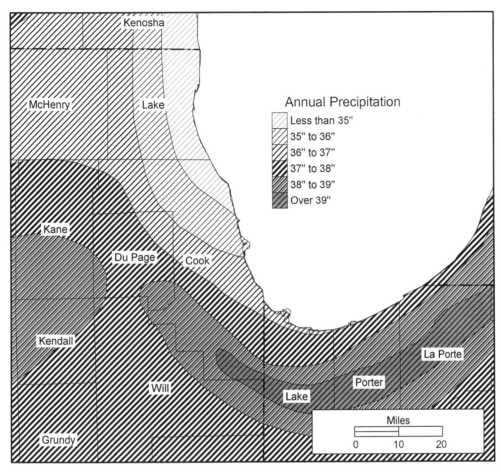

Fig. 4.3

Lake effect, topography, and the urban heat island have a complex influence on temperature patterns. The median date of last freezing (32°F) temperatures is a fairly early April 17th at Midway Airport, in the warm surroundings of the city. At Barrington, thirty-two miles northwest of downtown Chicago, the corresponding date is May 3rd, and in Wheaton, twenty-six miles west of the Loop, the median date of the last freezing temperatures is May 6th. Local influences on temperatures at individual weather stations have as great an effect on these values as do more systematic trends over the entire area. The same is true for the dates of the first freezing temperatures experienced in the autumn. On average, Midway Airport records its first freezing temperatures on Oct 23rd, compared with October 10th in both Wheaton and Aurora. The growing season is thus roughly a month longer near Midway Airport than it is in central DuPage County.

The map of modified growing degree days is calculated to show the optimal

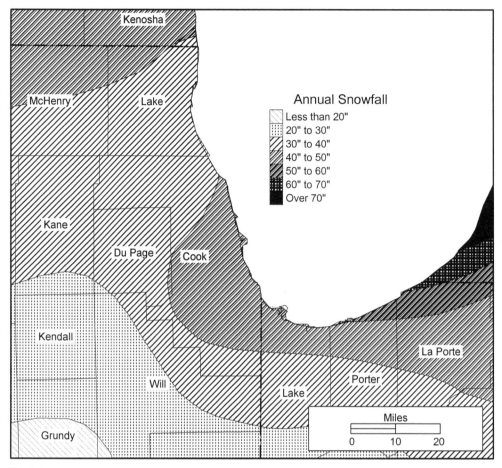

Fig. 4.4

growing season for crops such as corn (Fig. 4.5). Crops are not as successful if temperatures are too hot (over 86°F in the model) or too cool (below 50°F). Lake effect has a complex influence. Although it extends the growing season and reduces maximum summer temperatures, it also cools local temperatures. Across the border in Kenosha County, Wisconsin, conditions for the growth of crops such as corn are inferior compared with the warmer climate to the south, beyond Lake Michigan's influence.

Extreme weather events are remembered better than are average or typical conditions. Table 4.1 lists values of some extreme weather events recorded at O'Hare Airport beginning in 1958. Blistering heat, deep subzero cold, tropical deluges, and howling blizzards all have been recorded in Chicago, but they are quite rare. Location near the center of a large continental land mass produces marked seasonal variations in Chicago's temperature and precipitation, but it also offers protection against some of the more severe climatic extremes.

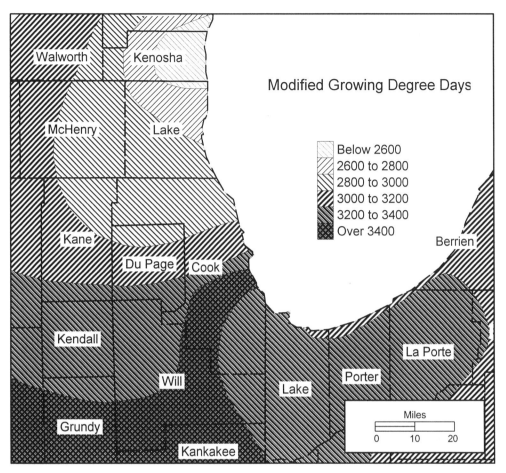

Fig. 4.5

Table 4.1: Extreme Weather Events at O'Hare International Airport, 1958–2005

EXTREME OF VALUE	DATE
Coldest temperature	27°F January 20, 1985
Heaviest rainfall	6.49" August 14, 1987
Hottest temperature	104°F June 20, 1988; July 13, 1995
Heaviest snowfall	18.6" January 2, 1999

Source: *Data provided by the Illinois State Climatologist's Office, a part of the Illinois State Water Survey.*

Chapter 5
CHICAGO'S LANDFORMS

Nearly all of the surface features in and around Chicago are the product of glaciation. The advancing glaciers eroded bedrock formations and ground the rock into boulders, cobbles, and sand. When the ice melted, streams issuing from beneath the glaciers laid down silt, sand, and clay, which were deposited in stream beds underneath the glacier or at its margins. After the last ice sheet disappeared about 13,500 years ago, the fine-grained deposits were subject to wind action and were blown to adjacent uplands.

Most prominent among the glacial formations is a series of low ridges, known as moraines, whose shape mimics the outline of Lake Michigan (Fig. 5.1). The lake's shape, in turn, matches that of a lobe of the continental ice sheet. Farthest from Lake Michigan is the Valparaiso moraine, which curves around the southern end of the lake from Indiana, through Illinois, and north to the Wisconsin border. East of the Valparaiso moraine is the Tinley moraine, which has the same shape bending around the southern end of Lake Michigan. Closer to the lake is a series of three smaller ridges, the Lake Border moraines, in Lake County and northern Cook County.

The moraines were produced by glacial advances during times of general retreat of the ice sheets. The Valparaiso moraine is older than the Tinley moraine which, in turn, is older than the Lake Border moraines. Lowlands between the moraines were filled with meltwater deposits emanating from the glaciers. Natural drainage of these features eventually resulted in stream formation. The parallel courses of the Fox, Des Plaines, and Chicago rivers resulted from the development of natural drainage on and between the moraines. Although the last glaciers disappeared from the Chicago area thousands of years ago, the deposits they left have been reworked by wind and water action ever since.

THE KANKAKEE ARCH AND THE CHICAGO OUTLET

Underlying geologic structures can have an important effect on surface topography, even though they are invisible. The Chicago area rests atop a buried geologic structure called the Kankakee Arch. The arch is composed of bedrock layers that are warped upward along a northwest-southeast axis that separates the Illinois Basin on the southwest from the Michigan Basin to the northeast. The Kankakee Arch is composed of dolomites that originated as shallow-water marine formations or as reefs during Silurian times (roughly 400 million years ago). These dolomites are crushed for road-building stone, although some rocks of similar age, such as the Waukesha dolomite, are quarried for building stone. The Silurian formations come closest to the surface at the apex of the arch in northern Will County.

Although the Kankakee Arch is a buried structure, it influences surface topography because it forms a bedrock sill that determines a maximum level for Lake Michigan. Lobes of the continental ice sheet occupied the Lake Michigan basin several times during the Pleistocene epoch (between two million and 10,000 years ago). When the climate warmed, melting took place around and underneath the ice sheet. Lake levels augmented by the melting glacier rose to the level of the bedrock sill and then spilled over into the Illinois Valley. The moraines, which consist of sand, pebbles, and cobbles, are fairly easily eroded. Rivers flowing away from the ice sheet's margins cut through the moraines, but were unable to cut through the bedrock sill, resulting in a ponding effect. The sill, known as the Chicago Outlet, thus determines a maximum size for Lake Michigan (Fig. 5.1).

The surrounding topography roughly resembles a saddle. Higher ground of the Valparaiso and Tinley moraines borders the Chicago Outlet to the northwest and southeast. Elevations on the moraines are 150-200 feet higher than the level of the Des Plaines River where it breaches the Chicago Outlet at Lemont. To the northeast lies the slightly lower elevations of Lake Michigan's margins, which were repeatedly transgressed by the lake's fluctuating level. To the southwest lies the Illinois Valley, all of which is lower in elevation than Lake Michigan. As geologist H. B. Willman remarked, "on the streets in Morris you are hardly aware that you are 75 feet lower than the surface of Lake Michigan" (Willman, 1971; 59). The Chicago Outlet has acted like a dam, holding the lake back until its level reaches a sufficient height, then allowing its excess to spill down the Illinois River.

GLACIAL LAKE CHICAGO

Chicago's topographic variation is subtle, but landform features do exist. In most cases, they can be identified in terms of variations in former lake levels. Glacial Lake Chicago is the name given to the high-water stages of Lake Michigan that existed

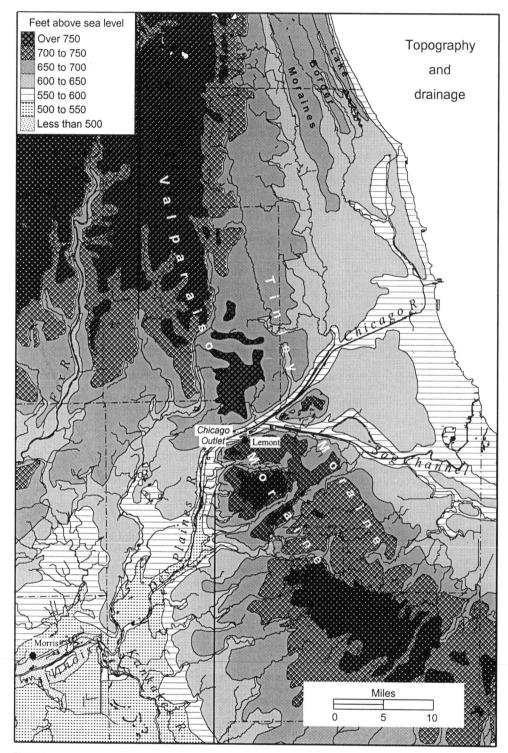

Feet above sea level
Over 750
700 to 750
650 to 700
600 to 650
550 to 600
500 to 550
Less than 500

Topography
and
drainage

Valparaiso

Tinley

Lake Border Moraines

Fox R.

Chicago R.

Chicago Outlet

Lemont

Moraine

Sag Channel

Des Plaines

Morris

Illinois

Miles
0   5   10

Fig. 5.1

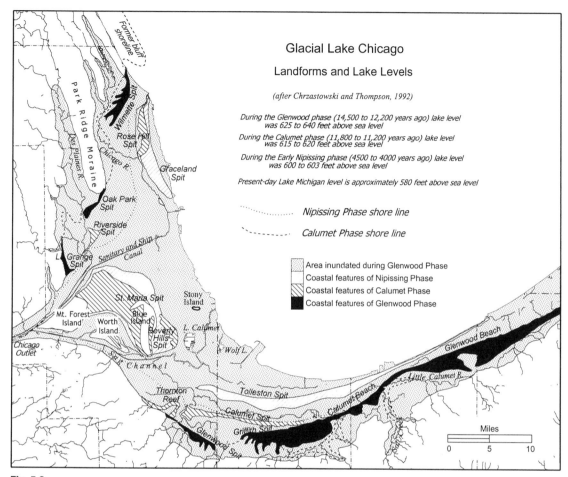

Fig. 5.2

during late glacial times (Fig. 5.2). The first, and largest, of the inundations was the Glenwood Phase, which took place between 14,500 and 12,200 years ago. This was followed by the Calumet phase (11,800 to 11,200 years ago) and the Early Nipissing phase (4,500 to 4,000) years ago. These high-water stands of Lake Michigan are not the only fluctuations in the lake's shoreline. Between 9,000 and 4,000 years ago, a low-water phase, called Lake Chippewa, saw a lake 350 feet below the present level. During the Lake Chippewa phase, hunting camps and other human habitations were constructed on dry ground well beyond the present limits of land.

Landforms produced during the Glenwood phase include a series of beaches and spits (narrowing points of land consisting of sand and gravel deposited by wave action and currents). The source of sand and gravel for these features probably was the eroding shorelines and bluffs along both sides of the lake. The Wilmette and Oak Park spits were built from materials eroded from the Lake border moraines to the north.

The arcuate shapes of the spits indicate that currents were directed toward the south. In similar fashion, eroding bluffs and beaches along the Michigan side of the lake provided materials for the Griffith and Glenwood spits in northwest Indiana.

Blue Island on Chicago's South Side has maximum elevations of over 665 feet. The small island stood above the waters of Lake Chicago during all of the high lake-level phases. Wave and current action directed from the north built spits on both the north (St. Maria spit) and east (Beverly Hills spit) sides of Blue Island. The St. Maria spit coincides today with the higher ground of St. Mary's and Evergreen cemeteries along 87th Street, between Kedzie and Crawford avenues. Prospect Avenue between 103rd and 111th streets outlines the shape of the Beverly Hills spit.

During the subsequent Calumet phase, all of the land between Blue Island and Mt. Forest Island remained above water level, as did Worth Island (present-day Holy Sepulchre Cemetery in the village of Worth). Because the Calumet-phase inundation was not as large, features resulting from it are found generally lakeward of the Glenwood spits and beaches. Rose Hill and Calumet spits are roughly four to five miles closer to Lake Michigan's present shoreline than the Glenwood-stage landforms.

A later resurgence of the lake's level, known as the Nipissing phase, came several thousand years after the Glenwood and Calumet highs. The western Great Lakes had been draining via the Nipissing River into the Ottawa River in Canada before that time; but, as the earth's crust rebounded after the weight of the glacial ice was removed, the erstwhile Nipissing outlet eventually rose above lake level. The Chicago Outlet briefly reemerged as one spillway after that time, but by about 4,500 years ago the St. Clair River at Port Huron, Michigan, emerged as the only outlet. As Lake Michigan's level rose to the Nipissing phase, approximately twenty feet above current lake level, the Graceland and Tolleston spits were formed. Stony Island, bounded by 91st and 94th streets between Stony Island and Colfax avenues, was an existing bedrock feature before glaciation, but it was above water level only during the Nipissing phase.

Gradual erosion of the Port Huron outlet led to a progressive lowering of lake levels until roughly 1,500 years ago, although dredging for navigation purposes has lowered it further in recent years. The last infilling (prior to shoreline changes caused by human activity) took place near Wolf Lake where the Illinois-Indiana border meets Lake Michigan. Materials carried by southward-moving currents along both the eastern and western shores eventually were deposited at Lake Michigan's southernmost extremity. As deposition and currents continued to fill and sculpt the southern end of the lake, the familiar smooth-arc contour from Chicago to Gary emerged. The Grand and Little Calumet rivers became drainage outlets as the land bordering the Tolleston Spit was left above lake level. Some relict beach lines within the city of Gary are no more than 500 years old. The process of infilling never was completed, however, with both Wolf Lake and Lake Calumet remaining as backwaters.

PREHISTORIC CHICAGO

Chicago's natural environment has undergone radical changes during the past 14,000 years. Climatic change led to glaciation, and that, in turn, produced variations in the water level and shape of Lake Michigan. All soils of the Chicago region are derived either from the moraines, wind-blown glacial deposits, or lake-bottom sediments laid down during this time. Natural vegetation changed in response as the glaciers retreated, with a nearly complete transformation of the plant cover, depending on whether a site was on a well-drained upland, a smooth middle slope, or a swampy piece of bottomland.

Climate was the ultimate cause of these cycles of landscape change. High-water phases of Lake Michigan took place under relatively moderate temperature conditions when glaciers were melting. The spits and beaches built on the lake's margins may have had a cover of tundra plants because summer temperatures were too cool for trees to thrive. Winters were not bitterly cold, however. Higher ground of the moraines eventually developed a cover of boreal (cold region) forest species, such as the tamarack, spruce, and balsam fir.

Climatic warming caused the Lake Michigan glacier to melt, which led to the Calumet phase of the lake. Continuation of the warming trend caused all of the continental glaciers to disappear from the area. With no ice blocking drainage to the north, plus an available lower-level outlet through the Ottawa River before crustal rebound took place, lakes Michigan and Huron shriveled in size. This was the Lake Chippewa phase of Lake Michigan, which coincided with a period of general climatic warming between 9,000 and 4,000 years ago.

Chicago's climate probably resembled that of Oklahoma or northern Texas during Lake Chippewa times. Grasses advanced into what had been forested lands. Upland sites became dry and even former wetlands at Lake Michigan's margins dried up. Human inhabitants of the Chicago region at that time would have shifted their hunting patterns from a focus on aquatic species and woodland mammals to herbivores of the grassland. Their patterns of habitation also would have shifted, away from what had once been sources of fresh water along streams. They likely moved toward the shoreline of Lake Michigan as it shrank northward. The landscape was almost totally transformed from a moist, forested one to a subhumid type of vegetation and a more open plant cover of grasses, small trees, and shrubs. Subsistence hunting and gathering activities of humans who lived there would have changed as well.

For all of these reasons, it makes little sense to speak of Chicago's "original" natural environment, at least if a fixed, unchanging set of conditions is envisioned. What was true in one period was replaced by different conditions in the next. Environmental changes have been both frequent and sweeping during the past 10,000 years, a time period when people were present to witness the change.

LANDFORM PATTERNS

Chicago has neither mountains nor high hills. The greatest peaks in elevation today are human-built embankments and landfills. But even though human activity has overshadowed nature in this respect, Chicago does have landform features—many of them named—that can be recognized fairly easily (Fig. 5.3).

Apart from the Indiana Dunes, the hilliest portion of the metropolitan area is the lake-border moraine complex in northern Cook and southern Lake counties. The Highland Park, Deerfield, and Park Ridge moraines are glacial depositional features that have been eroded over time. Along Lake Michigan, the Highland Park moraine is the basis for the line of irregular coastal bluffs, now steepened by erosion of their eastern flanks by wave action and gullied through a combination of natural erosion and human activities. It is a landscape that is almost devoid of level land. Relief features on the moraines decrease in height westward from the lake to their limit at the Des Plaines River.

Three forks of the North Branch of the Chicago River drain the landward side of these moraines. The sluggish, winding Skokie River between the Highland Park and Deerfield moraines was dammed during the 1930s to produce the artificial wetlands of the Skokie Lagoons. The "Skokie Valley Route," which was once a nickname for a line of the Chicago North Shore & Milwaukee Railroad that followed the Skokie River, is now occupied by Edens Expressway north of Lake Street, Wilmette. The Skokie River is the East Fork of the North Branch of the Chicago River. The Middle Fork, less than half a mile to the west, is a natural river, but it serves mainly a drainage ditch through residential neighborhoods.

The larger West Fork drains the area between the Deerfield and Park Ridge moraines. South of Willow Road it flows through wetlands adjacent to the former Glenview Naval Air Station, just west of Waukegan Road in Glenview and Northbrook. The West Fork joins the Middle Fork/Skokie River less than a half-mile east of the village of Golf, on Forest Preserve land just south of Golf Road. Rolling topography surrounds the junction of these two branches of the river in the scenic, privately owned setting of the Glenview Country Club. The area looks like few others in the Chicago region.

The Wilmette, Rosehill, and Graceland spits accumulated in that order during the Glenwood, Calumet, and Nipissing phases of Glacial Lake Chicago. Ridge

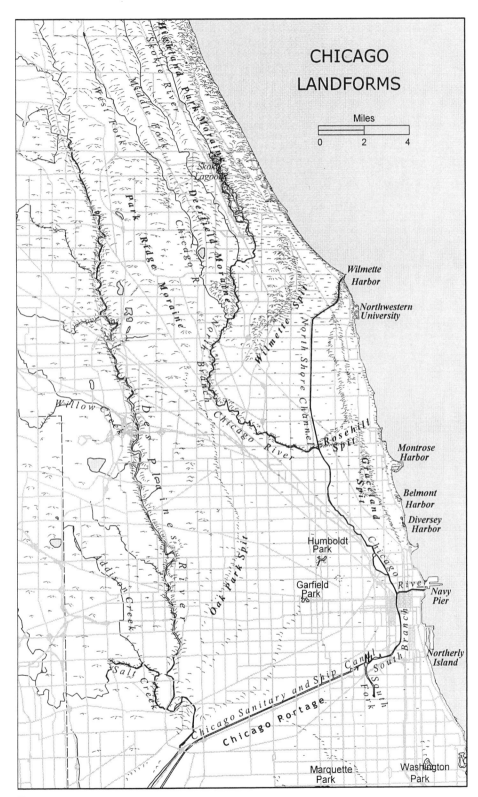

Fig. 5.3

Avenue and Gross Point Road follow the crest of the Wilmette spit for several miles into Chicago. Ridge Avenue in Evanston follows the trend of the Rosehill Spit. Ashland Avenue follows the Graceland spit from Devon Avenue to Irving Park Road. Rosehill and Graceland cemeteries occupy some of the hillier portions of the two spits and, of course, lend their names to these features.

Generally speaking, the city of Chicago was built on the flat bottom of Glacial Lake Chicago. Flatness of terrain has certain advantages, but it also creates severe problems of drainage, with which the city and its suburbs have been trying to cope for more than a century. The terrain is so flat that even the natural flow of streams has changed over time; that is, even prior to the era of human engineering projects.

The Wilmette, Rosehill, and Graceland spits diverted drainage southward and blocked the drainage of surface waters that otherwise might have flowed toward the lake. The Chicago River's abrupt southward turn at the tip of the Rosehill spit illustrates the role that even minor topographic features played in natural drainage. The present North and South branches of the Chicago River were doubtless, in some past time, parts of the same southward-flowing stream, a tributary of the Des Plaines River. The Chicago River's "natural" outlet near Navy Pier was natural only during part of this small stream's history, probably only within the past 3,000 years.

THE CHICAGO PORTAGE

Processes of coastal erosion and deposition continued even as Glacial Lake Chicago's level declined over the past 4,000 years. Erosion of the Graceland spit produced further deposition of sand to the south, which created beach ridges and dune features in the vicinity of Chicago's present downtown lakefront. The Chicago River was too small to breach this barrier, and its flow was directed south and west, making it a tributary of the Des Plaines.

By 3,000 years ago, however, the lakeshore had eroded landward to the extent that even a small stream such as the Chicago River was close enough and powerful enough to cut through the thin strand of beach and flow into the lake. It was at this time that the northern and southern portions of the Chicago River became its North and South branches (Chrzastowski and Thompson, 1992). While the North Branch remained relatively unchanged by this turn of events, the South Branch was reduced in volume of flow, and its upstream (western) end filled with silt. A roughly three-mile stretch, paralleling the present Chicago Sanitary and Ship Canal, from near Cicero Avenue to Stickney, became a dry stream bed except during periods of snowmelt and runoff. This necessitated the canoe portage for which Chicago became known.

The strategic value of the Chicago portage was known to the area's early inhabitants. In 1673, Father Jacques Marquette and Louis Jolliet traveled up the Illinois River and used the Chicago portage to enter Lake Michigan. The next year Father

Marquette returned and spent the winter of 1674-75 snowbound in a cabin at the portage. French traders used the Chicago portage as a canoe route between Lake Michigan and the Illinois River throughout the eighteenth century. The length of the portage varied with the year and season. In times of high water, the swampy former channel was navigable for much of its length, necessitating only a short portage. In dry weather, the traverse by land was longer. But it remained a drainage divide between the Great Lakes and the Mississippi River until the Illinois and Michigan (I&M) Canal was cut through it in 1848.

Several small streams flow eastward into the Des Plaines River, away from the crest of the Tinley Moraine. The largest of the tributaries, Salt Creek, is the main drainage of the lowland between the Tinley and Valparaiso moraines. These are sluggish streams during low-water periods, but they can fill quickly after a period of unusually abundant rainfall and during spring snowmelt. They are especially flood-prone because of the rapid drainage created by the many storm-sewer systems that now feed them.

THE INDIANA DUNES

The Indiana Dunes are the most striking topographic feature of the Chicago region (Fig. 5.4). A strip of sandy dunes begins at the southern tip of Lake Michigan and extends more than 200 miles along Michigan's side of the lake. Location of the dunes

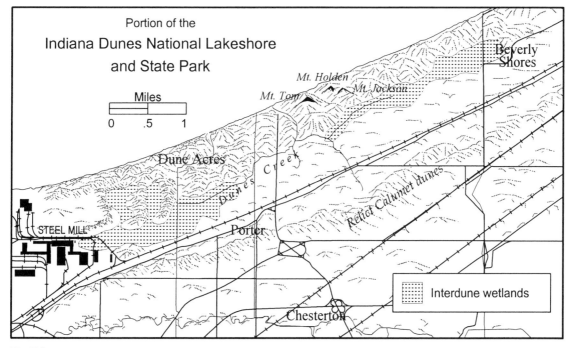

Fig. 5.4

is a key to their origin. Unlike the sand dunes of a desert, which form and reform over vast areas and which assume a variety of wind-sculpted shapes, Lake Michigan's dunes have been created in place and have migrated only a short distance from their source of sand.

Winds blowing down the length of Lake Michigan, or from west to east across the lake, exert considerable erosive force on sandy beach deposits. Sand is removed from the lake's immediate margins in a process known as deflation, which leaves a concave lakeward slope. Trees and other vegetation considerably slow the sand's landward migration and create a stable footing against which the sand accumulates. Individual dunes, such as Mt. Holden in the heart of the Indiana Dunes, rise 180 feet above the nearby lakeshore. The bounding vegetation, however, confines dune development to a strip little more than a half-mile in width.

The same process created coastal dunes at the various margins of Glacial Lake Chicago. Dunes built during the Calumet phase form a ridge about a mile inland from the present lakeshore. They rest on a surface that is approximately 620 feet above sea level, corresponding to the Calumet-phase water line. These older, relict dunes now have a thick forest cover and are no longer active, although removal of the vegetation allows wind erosion to resume. In modern times, the lowlands between parallel ridges of dunes offered smooth, natural corridors for railroads and highways to follow.

The wind's reworking of massive sand deposits blocked the natural drainage outlets of streams entering the lake. The result was a strip of wetlands between the dune formations. Sandy soils drain rapidly, whereas marshy and swampy soils typically remain waterlogged much of the year. Environments so different but so close to one another offer many varieties of plant habitat. More than a century ago, plant ecologist Henry C. Cowles of the University of Chicago recognized the diversity of flora within the dunes area and began to study it systematically. His observation of the succession of plant species occupying a site over time led to influential theories about vegetation succession (Cowles, 1901).

Unsuitable for agriculture, the dunes were bypassed by human activity for many years. The bunching of transportation routes near the lake made the entire area accessible, however, and some lakefront properties in the dunes were developed for recreation. Always a relatively wild area, but within sight of the city, the dunes were subject to various development efforts. Local citizens, including landscape architect Jens Jensen, organized to prevent destruction of the dune landscape by the early decades of the twentieth century. Preservationists faced their greatest challenge in the 1960s when steel companies made plans to develop extensive tracts of dune land for industrial purposes.

Best known of those who championed preservation efforts was Illinois Senator Paul C. Douglas. Local leaders enlisted Douglas's aid to get a portion set aside as the Indiana Dunes National Lakeshore in 1976. Other dune areas are included within the Indiana Dunes State Park. Steel mills were built on some dune land, however, with millions of cubic yards of sand excavated and barged away. Some of the sand went north to Evanston where Northwestern University dumped it in Lake Michigan, creating a strip of new land for campus expansion in the early 1970s. The Indiana Dunes retains its unusual mixture of uses, ranging from heavy industry, to recreational homes, to ecological laboratory, all within a few miles of one another.

Chapter 6
RIVER BASINS AND DRAINAGE

THE UPPER ILLINOIS RIVER BASIN is defined as the area drained by rivers and creeks emptying into the Illinois River upstream from the city of Ottawa (Fig. 6.1). Total area of the Upper Illinois Basin is 11,000 square miles, of which 10,000 is included in three basins: the Kankakee River (5,240 square miles); the Fox River (2,577 square miles); and the combined Des Plaines and Chicago rivers (2,183 square miles). Another 1,000 square miles comes from small basins between Morris and Ottawa. The Illinois River's starting point is the confluence of the Des Plaines and Kankakee rivers a few miles upstream from Morris. Only a narrow fringe of land in the city of Chicago, where surface runoff would drain directly into Lake Michigan if it did not enter the city's storm-sewer system, is excluded from the Upper Illinois Basin.

Natural history of the Kankakee, Fox, and Des Plaines-Chicago basins continues the story of environmental change that began with glaciation and the resulting fluctuations in Lake Chicago/Lake Chippewa. All three river basins were the sites of native villages and seasonal camps, and all three were exploited by Euro-American settlers who began arriving at Chicago in sizable numbers by the late 1830s.

KANKAKEE BASIN

The Kankakee has the largest drainage basin of any of the rivers near Chicago (Fig. 6.2). It is also perhaps the least known. Just as the Kankakee Arch's upward bend provides the bedrock sill that defines a maximum level for Lake Michigan, so does the same structure limit the gradient on the Kankakee River. The Kankakee was once a slow, meandering stream that drained westward from the undulating, glacial topography south of the Valparaiso moraine in Indiana. At Momence, just west of the Illinois-Indiana border, Silurian rocks of the Kankakee Arch are close enough to the surface to influence drainage. Unable to erode the underlying bedrock sill, the

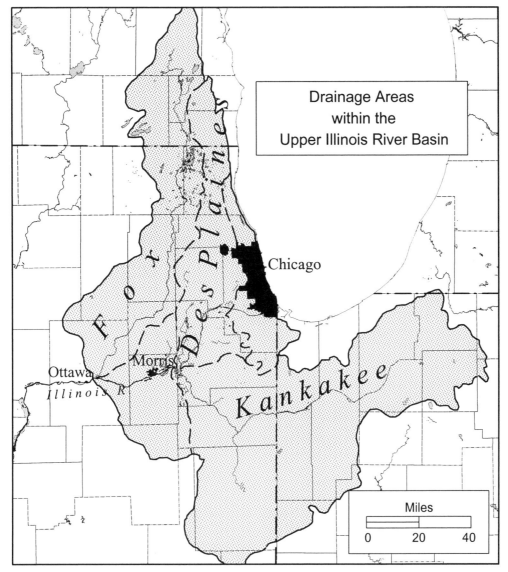

Fig. 6.1

slow-flowing Kankakee formed a series of tight meanders across its floodplain in Indiana. A sinuous river channel perhaps 250 miles long wound along a straight-line axis of roughly ninety miles from South Bend to Momence. Much of the Kankakee River's upper basin had no well-defined pattern of drainage until ditches were constructed in the early twentieth century. The geometric drainage pattern of human-built ditches in Indiana today contrasts sharply with the Kankakee's natural drainage pattern in Illinois. The Iroquois River, the Kankakee's principal tributary, enters the main river in the city of Kankakee.

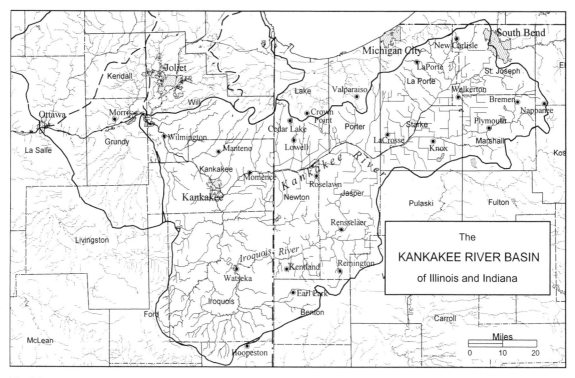

Fig. 6.2

Despite its meandering ways, the Kankakee did offer a canoe route from the future site of South Bend to the Illinois River. At South Bend, only a short portage was necessary to reach the St. Joseph River, which discharges into Lake Michigan. Traders and exploring parties used this route, rather than the Chicago portage, at various times during the fur trade era of the eighteenth century. Apart from this use, which ended well before Euro-American settlers began moving into northern Indiana, the Kankakee basin was bypassed. Its thousands of acres of marshy wetlands alternated with occasional swells of higher ground which supported heavy timber (Fig. 6.3). Natural levees along the Kankakee's banks also had dense stands of hardwoods. It was a paradise for duck hunters, who built more than a dozen hunting lodges in the Kankakee Marsh after the 1880s.

Without artificial drainage, agriculture was severely limited to scattered upland ridges and knolls. Seasonal floods prevented most permanent human habitations elsewhere. The construction of drainage ditches was accompanied by a massive channel-straightening project begun by the State of Indiana in 1896. Because the Kankakee's flow was unimpeded west of Momence, the State of Illinois did not participate in the channel-straightening project, although drainage for agriculture took

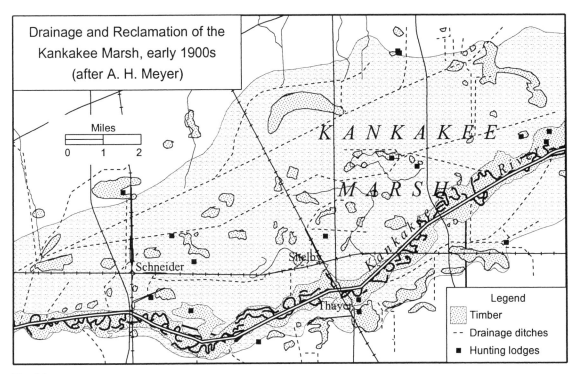

Fig. 6.3

place on the Illinois side as well. Between 1896 and 1920, Indiana's portion of the Kankakee Basin was transformed from marshland and swamp into cropland and pasture, and thus it has remained. Strands of backswamp and hundreds of cutoff meanders remain from the river's old course, but the Kankakee itself has been a human-built ditch for more than a century.

Poor natural drainage was not confined to the Kankakee River basin. All of the prairies south and west of the Valparaiso moraine in Indiana and Illinois were classified as wet prairies, meaning they were subject to seasonal standing water and hence were unreliable for crops such as corn and wheat. The first use that Euro-American settlers made of the wet prairies was to pasture livestock. The Grand Prairie of Illinois and Indiana covers roughly 50,000 square miles south of the Illinois River and overlaps the western half of the Kankakee River basin. The Grand Prairie was excellent range for cattle prior to the 1870s when farms were few in number and widely scattered. Thousands of acres were fenced and used as seasonal range for young cattle, much as the Great Plains was used in later years.

Southern portions of the Kankakee basin, especially the area drained by the Iroquois River, were home to several large-scale attempts to raise cattle on the prairies.

Although the wet prairie could be used as cattle pasture, drained prairies converted to cropland could be used to raise feed crops to fatten cattle for market. Drainage efforts increased during the 1880s. By 1900, the Grand Prairie had been drained, its lands had been turned into productive cornfields, and Chicago was the world's leading corn market.

FOX BASIN

The Fox River Basin is only half the size of the Kankakee, but it has had a larger human population ever since Euro-American settlement began in the 1830s (Fig. 6.4). The Fox drains the area between the Des Plaines River on the east and the Rock River (a tributary to the Mississippi) on the west. The name Fox, while perhaps attributable to the historic presence of the well-known fur-bearing animal, is said to have been given in reference to the river's meandering course, zig-zagging across the landscape with many abrupt changes in direction, as is the habit of a fox. The river's many meanders are a result of its gentle gradient across a glaciated land dotted with ponds, lakes, and marshes. Although much of the Fox basin was poorly drained in its natural state, it was part of a landscape that included many smooth uplands. The uplands, with a vegetation cover of oak woodland and grass, were easy to bring into cultivation by the Yankee farmers who came to northern Illinois and southern Wisconsin in search of good land in the 1830s. Nearly all of them sowed wheat or other small grains on the first lands they cleared. They were rewarded with harvests far larger than they had known in western New York State.

Some wheat was shipped down the Fox River when water levels were high, although most farmers hauled grain overland by wagon to one of the lake ports. Chicago was an important shipping center, but so were Waukegan, Kenosha, Racine, and Milwaukee. The early wheat crops of the Fox River valley also stimulated the construction of grist mills and flouring mills, run by water power, along the Fox and some of its tributaries. Both Aurora and Elgin, on the Fox River in Illinois, were early milling centers, as were Burlington and Waukesha in Wisconsin. Small-scale industrial activities associated with milling local crops gave way to other types of manufacturing industries once wheat went into decline, but the industrial nature of Fox Valley cities had been established.

Continued wheat farming eventually depleted the land of its long-stored natural fertility, and by the 1860s the wheat frontier began to move west. Production costs were lower on newly settled frontiers, first in western Wisconsin, then in Minnesota, and finally in the Dakotas. The Kankakee basin and its surroundings evolved a corn-livestock economy while the Fox Valley shifted from wheat farming to dairying. These two systems of Midwestern agriculture tended to remain distinct in later

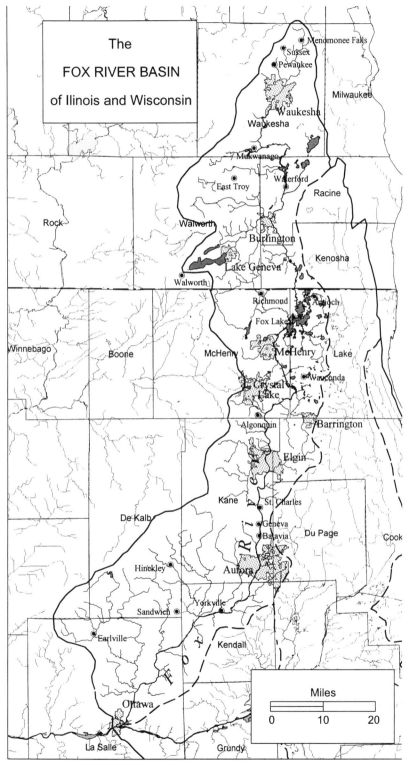

The

FOX RIVER BASIN

of Ilinois and Wisconsin

Fig. 6.4

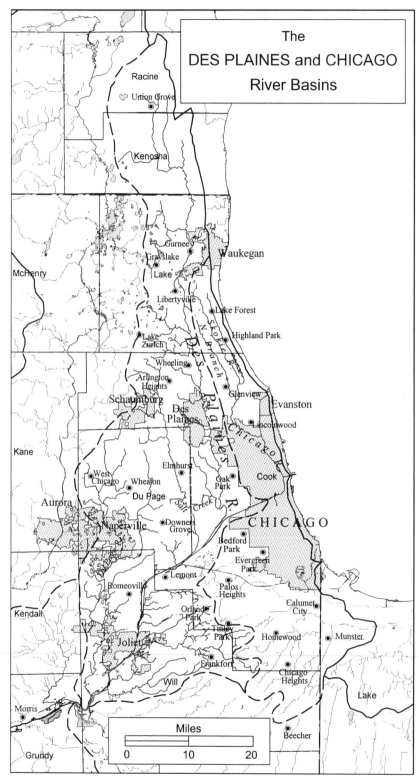

The
DES PLAINES and CHICAGO
River Basins

Racine
Union Grove
Kenosha
McHenry
Gurnee
Grayslake
Lake
Waukegan
Libertyville
Lake Forest
Lake Zurich
Highland Park
Wheeling
Arlington Heights
Schaumburg
Glenview
Evanston
Des Plaines
Lincolnwood
Kane
Elmhurst
West Chicago
Wheaton
Oak Park
Cook
Du Page
Salt Creek
Aurora
Naperville
Downers Grove
CHICAGO
Bedford Park
Evergreen Park
Lemont
Romeoville
Palos Heights
Kendall
Calumet City
Orland Park
Timley Park
Homewood
Munster
Joliet
Frankfort
Will
Chicago Heights
Lake
Morris
Beecher
Grundy

Miles
0   10   20

Fig. 6.5

years as both moved westward: wheat farming to the northwest and corn-livestock agriculture to the west. Chicago gained from its proximity to both agricultural modes, and the city became an important marketing point and processing center for grain and livestock. Intensive dairy farming was practiced in the Fox River valley from the late decades of the nineteenth century until the late twentieth, when production cost increases favored the migration of dairying northward in Wisconsin.

The many lakes within the Fox basin were attractive to vacationers even in Chicago's early days. Creation of frequent rail commuter service in the early decades of the twentieth century made it possible for people of some means to live in rural settings yet still work in the city. Many communities in the Fox basin—including a few in Wisconsin, such as Walworth and Lake Geneva—evolved exurban economies based on commuting to Chicago. Recreation developments, second homes, and permanent commuters thus all mingled in the lake-dotted landscape of the Fox basin.

DES PLAINES AND CHICAGO RIVER BASINS

The Des Plaines River's drainage area resembles that of the Fox. Both rivers reach well into Wisconsin. They flow through a landscape of lakes, marshes, and wooded uplands and eventually empty into the Illinois River. The Des Plaines River's two main tributaries are Salt Creek, which enters the main channel at the community of Riverside, and the DuPage River, which flows into the Des Plaines near its confluence with the Kankakee (Fig. 6.5).

The Des Plaines and the Chicago rivers have distinct natural drainage areas, separated in the north by the higher ground of the Park Ridge moraine. For hundreds of years prior to the twentieth century, they were merely adjacent; for the past century, the Chicago River's flow has been diverted into the Des Plaines and thereby made tributary to the Mississippi River.

The Des Plaines River has a well-developed drainage network west of the Valparaiso moraine. Between the high ground of the moraine and Lake Michigan, however, the land was poorly drained before ditching took place in the early twentieth century. Thousands of years ago, the lowland through which the Cal-Sag Channel now passes drained through the Chicago Outlet during high-water periods; but no natural-flowing stream continued drainage along that route in later years. The "Sag" name is derived from Saganashkee, which was a swamp or marsh until 1922 (Fig. 6.6). The Grand Calumet and Little Calumet rivers originated in Indiana, as they do today, but they flowed via Lake Calumet or through Wolf Lake, entering Lake Michigan through the large marshy area along the Illinois-Indiana border.

In 1922, the Chicago Sanitary District completed the Cal-Sag Channel (Fig. 6.7). The Calumet rivers' natural outlet was blocked at the Thomas J. O'Brien Lock and

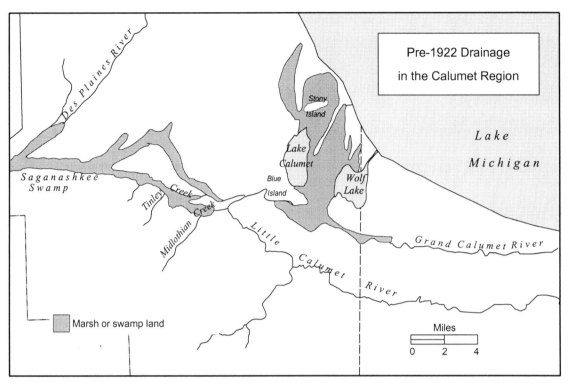

Fig. 6.6

Dam at 134th Street and Torrence Avenue. The Little Calumet River was dredged, as was a new waterway through the Saganashkee Swamp, creating a new outlet for drainage to the west. Barges traveling up the Illinois and Des Plaines rivers were able to reach Lake Michigan by using the O'Brien Lock, stepping up to the slightly higher elevation of the lake.

Construction of the Cal-Sag Channel was the third major modification of local drainage patterns undertaken by the Chicago Metropolitan Sanitary District. The Sanitary and Ship Canal, completed in 1900, connected the Chicago River with the Des Plaines, paralleling the already existing Illinois and Michigan Canal. In 1907, the North Shore Channel was built to connect the city's North Side and several northern suburbs with the Chicago River.

The major purpose of both of these channels was to prevent discharge of waste water into Lake Michigan. Treatment plants were built in connection with the drainage channels to treat raw sewage before it was released down the Des Plaines River and into the Illinois. Capacity of the plants was exceeded during periods of heavy stormwater runoff. The surging runoff mixed with untreated sewage from the storm sewers, bypassing the treatment plants.

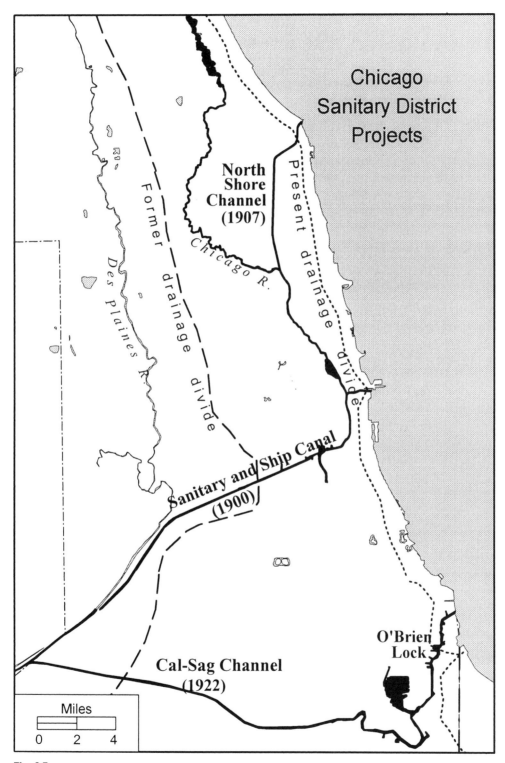

Chicago
Sanitary District
Projects

North
Shore
Channel
(1907)

Present drainage divide

Former drainage divide

Des Plaines R.

Chicago R.

Sanitary and Ship Canal
(1900)

Cal-Sag Channel
(1922)

O'Brien
Lock

Miles

0   2   4

Fig. 6.7

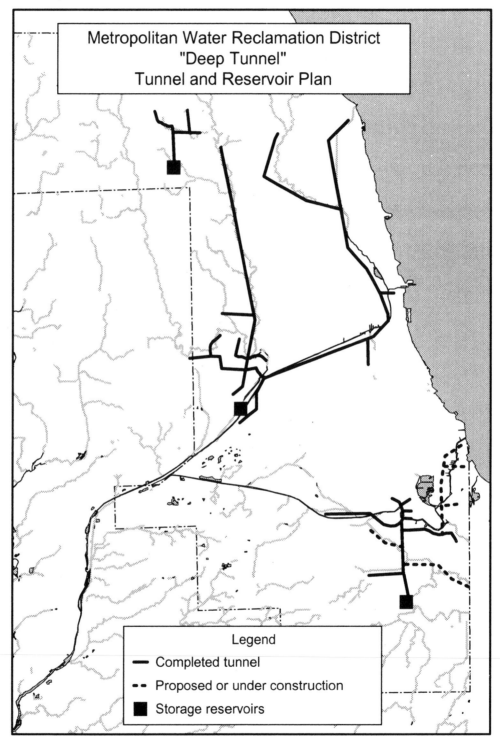

Metropolitan Water Reclamation District
"Deep Tunnel"
Tunnel and Reservoir Plan

Legend
— Completed tunnel
-- Proposed or under construction
■ Storage reservoirs

Fig. 6.8

In the 1970s, a new Tunnel and Reservoir Plan was begun (Fig. 6.8). Popularly known as "Deep Tunnel," the first link in the new system was a thirty-one-mile tunnel extending from Wilmette to the Southwest Side of Chicago. Other tunnel systems were constructed near O'Hare Airport, parallel with the Des Plaines River and near Lake Calumet. The new underground pattern of drainage adds capacity to the existing surface system and along the familiar drainage routes that have functioned, in one capacity or another, for thousands of years.

The truncated Little Calumet and Grand Calumet rivers no longer enter Lake Michigan through the swampy outlet north of Lake Calumet. Both rivers enter the lake farther to the east in Indiana where short channels divert their flow. Marshy or swampy wetland areas typically are a casualty of drainage and stream channeling projects, and this is true of the Calumet area, in particular. Large areas of seasonal standing water still are found near Lake Michigan's southern shore, but the unique marsh environments—a mixture of land and water—have largely disappeared.

Part III

HISTORICAL PATTERNS

Chapter 7
ORIGINS OF THE CITY

Chicago's site was not optimal for a city because of drainage problems associated with the flat terrain. To use a familiar distinction, Chicago's site was only fair, but its situation—on the divide between two great waterway systems of eastern North America—made it almost inevitable that human activities would congregate there. The specific condition that made settlement likely in early historic times was the natives' and French traders' use of canoes, which could be portaged across the low drainage divide. Portage sites had to be defended against possible interference from rival traders. They were likely places for the barter and exchange of goods to take place and attracted various other forms of economic, governmental, or military activity. Early Chicago was a portage, trading post, military fort, and place of congregation at times when treaties were being arranged.

EARLY WATER ROUTES

Chicago's portage was not unique, even in the lower Great Lakes region. When Father Jacques Marquette and Louis Jolliet made their first entrance into the Mississippi River system, they went farther north by way of Green Bay and the Fox-Wisconsin waterway. In the spring of 1673, they turned back on their voyage down the Mississippi and took a shortcut, well known to the native people, that led up the Illinois River to the Des Plaines and then to the short portage to the Chicago River and into Lake Michigan. Directly across Lake Michigan from Chicago is the mouth of the St. Joseph River. In 1679, French explorers ascended the St. Joseph to its headwaters near South Bend, Indiana, and then followed the Kankakee River westward to its confluence with the Des Plaines and the beginning of the Illinois River. All three routes—via Green Bay, Chicago, and St. Joseph—were used by various explorers and trading parties during the late seventeenth and eighteenth centuries. A fourth route, via the Maumee River with a portage to the Wabash River at Fort Wayne, Indiana, also was used (Fig. 7.1).

Each of the four routes had advantages and problems. The Maumee/Wabash route was shortest because it avoided the long detour around the lower peninsula of Michigan, but the portage between the two rivers was lengthy, and low-water problems plagued the upper reaches of both the Maumee and the Wabash. The Fox-Wisconsin route, being farther north, was open fewer months of the year. The Wisconsin River portion was easy, but the upper Fox was a winding, shallow stream.

The Illinois River offered the fastest route to the Mississippi. A broad river of substantial size, it was navigable over a long season and flowed through a country that had a large native population. Of the two routes to reach the Illinois River from Lake Michigan, the St. Joseph/Kankakee was longer and slower. This left the Des Plaines River route. Traders favored it as a means of travel between Montreal in Quebec and the mouth of the Mississippi south of New Orleans.

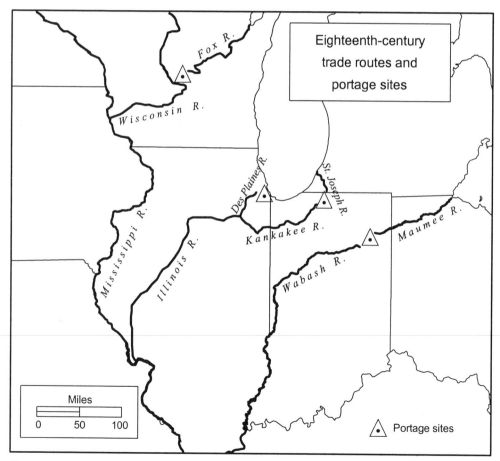

Fig. 7.1

French traders and missionaries are recorded as having made several passages through Chicago in the years following Jolliet and Marquette's first visit in 1673. No permanent developments were to follow for some time, however, because of a series of armed struggles between the natives and the Europeans and between the two European powers, France and Great Britain. The Fox Indians of Wisconsin turned against the French, with whom they had once been allies, and this diverted trade away from any route to the Mississippi via Lake Michigan. The first half of the eighteenth century saw little activity at Chicago. During the Seven Years War (1756–63) between France and Great Britain, the Chicago route had renewed activity as a result of military operations—a prelude to the fall of France in the early 1760s and the subsequent British takeover of the lands the French had claimed, which included the entire drainage area of the Great Lakes.

THE FORT DEARBORN ERA

The American Revolution (1775–83) followed little more than a decade after the British had forced out the French. Defeat of the British brought the western Great Lakes region under American control, but the frontier was dangerous, and those who chose to live at outposts such as Chicago were subject to the hostilities that lingered among the native people, who remained loyal to the British. The 1795 Treaty of Greenville established some American authority in the western settlements. The new nation began to construct military forts at strategic locations, one of which was Fort Dearborn, which was built on high ground behind the beach line of Lake Michigan, just south of the mouth of the Chicago River, in 1803 (Fig. 7.2).

The construction of Fort Dearborn marked another step in the development of what would become the city of Chicago. A six-mile-square tract of land on which the fort was built had been secured via a native land cession in 1795. Fort Dearborn had only a small garrison, but attached to it were an Indian agency and a factory (fur trading post). It was a mixed-purpose, fortified settlement of the sort that was common on the western frontier during the fur-trade era.

Among the men often mentioned in accounts of Chicago at this time are Jean Baptiste Point Sable (DuSable) and Anton Ouilmette. DuSable, of African and French ancestry, was Chicago's first long-term resident. He lived at Chicago during the last two decades of the eighteenth century, before Fort Dearborn was built. The British distrusted him, as they did nearly all French traders, yet DuSable prospered as a trader after the period of French influence had passed. Ouilmette, roughly a contemporary of DuSable's, was a French Métis (an individual of mixed blood) from Canada who similarly carried on an active trade with the native inhabitants of the region. Both men were engaged in farming. They planted corn and maintained herds of livestock to support the local trade. DuSable left Chicago in 1800, but Ouil-

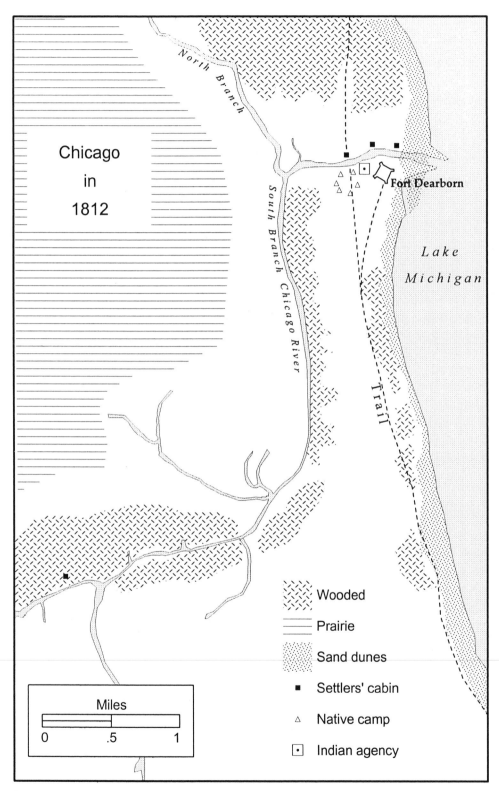

Chicago
in
1812

North Branch

South Branch Chicago River

Fort Dearborn

Trail

*Lake
Michigan*

Wooded

Prairie

Sand dunes

■ Settlers' cabin

△ Native camp

⊡ Indian agency

Miles

0   .5   1

Fig. 7.2

mette remained and eventually received a grant of land to the north of the city, from which the present-day village of Wilmette takes its name.

DuSable and Ouilmette were transitional figures in every respect. Of partial French ancestry, they were middlemen between trading parties from the East and the local native populations. They traded goods, raised food, served as interpreters, helped settle disputes, and were relied upon by all parties as sources of information about the locality. Their many roles were temporary, reflecting the impermanence of conditions and the necessity for adaptation to outside forces. Chicago was to continue in this role as a small, informal settlement for some time.

The War of 1812 exacted a terrible toll on Fort Dearborn and its inhabitants. Although their chiefs desired peace, many Indians living near Chicago sought to continue the hostilities that had most recently been in evidence at the 1811 Battle of Tippecanoe in northern Indiana. Allied with the British, the natives began to attack isolated frontier settlements. A series of military setbacks in Canada led the American authorities to order a retreat from various frontier outposts, including Fort Dearborn, in the summer of 1812. Abandoning the fort, the soldiers, other local citizens, and some two dozen women and children set off on the trek east to Fort Wayne on August 15, 1812. Vastly outnumbered, the small group was ambushed as they passed by the dunes along the lakeshore. Those who were not killed outright were captured and put to death or else later perished in captivity. Intervention of U.S. authorities led to the natives' release of several officers in the ill-fated Fort Dearborn garrison, who survived and later told of the events.

The original Fort Dearborn lay in ruins, but, once peace with Great Britain was established in 1816, the matter of national sovereignty was settled. There would be no more wars over territory in the Great Lakes region. Under Army orders, a new Fort Dearborn was completed in May 1817. Trading activities resumed, although the pace of activity was slow and for a time the military abandoned Fort Dearborn. In treaty after treaty, some of which were negotiated at Chicago, the native peoples of northern Illinois and southern Wisconsin gradually lost their lands near Chicago and were forced to reservations west of the Mississippi River. Fort Dearborn finally was closed for good in 1833. The frontier outpost phase of Chicago's history thus came to an end.

PLANS FOR A CANAL

Even while the second Fort Dearborn was under construction, Major Stephen Long visited Chicago to make preliminary surveys for a canal that would connect Lake Michigan with the Illinois River. Internal improvements such as canals were under federal jurisdiction, although they were matters of intense state and local interest as well. When Illinois was admitted to statehood in 1818, the plan for a canal linking the Great Lakes and the Mississippi River was a major argument in favor of including what are now the northern counties of Illinois in the new state. The canal would not be completed for another thirty years, but projections of the economic activity it would bring attracted speculative investment and led to population growth at Chicago.

The desirability of linking the Great Lakes with the Mississippi River was of greater interest to the nation as a whole, perhaps, than to Chicago. There was little commerce on the Great Lakes until the Erie Canal was completed in 1825. While Chicago had access to the East via water, what it crucially lacked was any access downriver to the South and West, which was where nearly all of the settlement growth in the Middle West took place before the 1830s.

Chicago was peripheral to Illinois because it had no connection to the Illinois River. The city played a minor role in the state's development during the early decades of the nineteenth century. Settlers from Kentucky streamed into the Sangamon country near Springfield, Illinois, during the 1820s, where they established corn and livestock agriculture. Settlement spread up the Wabash and Mississippi river valleys at the same time. The lead mines around Galena in northwestern Illinois were in active production. Chicago was peripheral to all of these developments and gained little in population or economic growth as a result of their presence.

One indication of Chicago's remoteness was its reliance on outside sources of supply for its food. Drovers walked herds of cattle north from the Wabash Valley to sell in Chicago. They approached overland from the south, roughly along a route that became one of the city's streets, hence the name Wabash Avenue. In time, Chicago would become the great market, processing center, and shipment point for grain and meat in the Middle West, but, until it had transportation access to its hinterland in the nearby Illinois Valley and Grand Prairie, it could play no such role.

CHICAGO'S FIRST PLAT

The presence of economic activity was not a prerequisite for calling a new town into being. By 1830, the common practice in town-founding was to plat in advance of settlement. A tract of land was acquired, streets were surveyed, building lots were demarcated, and the surveyor's plat was filed with the county recorder of deeds. The plat map was used to sell real estate, often before much settlement or economic activity had taken place. This was the manner in which Chicago was created.

In 1827, the federal government granted land to the State of Illinois for construction of a canal to link Lake Michigan with the Illinois River. Terms of the grant awarded the state alternate (odd-numbered) sections of land (one-mile square, or 640 acres) within a distance of five miles on either side of the proposed route. The I&M Canal commissioners authorized surveys of several town plats, including those for

Chicago and Ottawa, at the eastern and western ends of the canal's route. Sales of lots in the new towns were made in order to cover expenses of the survey. Chicago's first plat, made by surveyor James Thompson, was filed in 1830. The next year, twenty-four lots were donated to the newly formed Cook County for purposes of constructing public buildings, although sixteen of the lots soon passed into private hands.

Surveyors' town plats had to look like credible towns in order to stimulate investors' interest. Most, like Chicago's, followed a simple grid-pattern design. The initial plat covered the south half of section 9, township 39 north, range 14 east, on one of the I&M Canal grant's odd-numbered sections (Fig. 7.3). The plat was centered on the junction of the North and South branches of the Chicago River. By 1834,

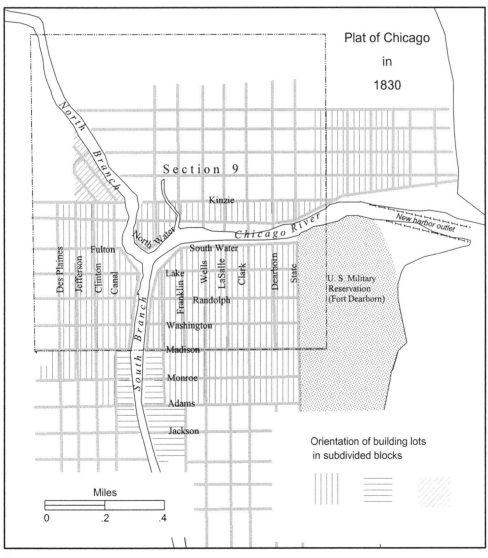

Fig. 7.3

the initial plat had been enlarged several times, with new additions in the even-numbered (public land) sections on all sides, no doubt to accommodate the growth that was expected.

Platting a town ahead of settlement had both the advantage and disadvantage of the speculation in real estate that typically followed. Speculators bought, sold, and traded urban real estate before it was occupied, eventually establishing a price for town property that reflected the public's overall assessment of how successful the place would become. If the speculative phase was brief and was redeemed by the appearance of new economic activities, it served a useful purpose; but if actual growth was slow in coming and speculation merely led to more of the same, the eventual result was a sudden, drastic reduction in lot prices that emerged from the turmoil of panic selling. The latter was to be Chicago's fate, although the bubble was a long time in bursting.

What kind of community did the platters intend Chicago to be? There can be little doubt that they believed the Chicago River would be the community's focus. Centering the plat around the junction of the North and South branches of the Chicago River maximized the number of feet of river frontage, guaranteeing a large number of business sites with direct access to the river. The pattern of building lots in the subdivided blocks reinforces the impression that river access was highly desirable. The main business streets, including Lake, Randolph, Washington, and Madison, ran east and west, down to the riverbank. The initial plat indicated no development on the lakeshore, nor would that have been possible in 1830, given the lingering presence of Fort Dearborn.

The plat effectively divided Chicago into three riverside communities. It was an early sign of the division and partition that would segment the city into sectors, wedges, and "sides" from that time forward. Chicago was a river town, although the river in section 9 was actually three small, sluggish stream segments that had partially blocked drainage into Lake Michigan half a mile to the east and were navigable no more than a few miles inland, either north or south. It was a somewhat inauspicious beginning for a site that would soon become and long reign as the transportation hub of the mid-continent region.

TRANSPORTATION IMPROVEMENTS

The streams were small, but the surrounding topography made them difficult to ford. The first bridge connecting the north and south banks of the main channel was built by soldiers at Fort Dearborn in 1832. After the military had left, it was up to the city to establish bridges or ferries, but there were difficulties. Ship captains entering the Chicago harbor preferred ferries because they did not obstruct the channel.

A bridge at Dearborn Street, constructed in 1834, was demolished in 1839 because it was a hazard to navigation, and a ferry was substituted. Bridges across the main channel at Clark Street and across the South Branch at Randolph Street finally were constructed in the mid-1840s.

The slow pace of bridge construction—even more the removal of bridges and substitution of ferries—reflects a lack of demand for access across the river. The south and east banks remained the favored sites for business growth into the mid-1840s, and both could be reached overland from the south, which was the major direction of trade. One of Chicago's business pioneers, William Butler Ogden, owned property north of the river. He observed that his section of the city was developing slowly because of the lack of bridge crossings (Pierce, 1937; 341). The same could have been said of the even less accessible west bank.

If the city's accommodation to the rivers in its midst was uncertain, it was even less successful dealing with the harbor problem. In 1817, Major Long had recommended constructing embankments into the lake and digging a channel between them as a way of cutting through the natural accumulation of sand at the Chicago River's mouth. Long's recommendations were followed, although the new harbor outlet soon filled with sand, deposited by the ever-present longshore current.

The federal government had control of the rivers and harbors, but government appropriations to improve Chicago's harbor were slow in coming. The city eventually had to take its own measures to correct the problem. Great Lakes ships often stayed offshore and transloaded cargoes to smaller vessels that could navigate the river's mouth. By 1848, when the I&M Canal was completed and the first railroad began operations, Chicago's harbor had been improved enough to make the city the lake port it was intended to be.

In nearly every respect, Chicago was not much of a success in its early years. More than a century elapsed from the time of Jolliet and Marquette's initial visit until the first Fort Dearborn was built. Although the potential of Chicago's site was acknowledged widely, there was little development until conditions, both near and far, had improved. The value of the Great Lakes for shipping increased markedly after the Erie Canal was completed in 1825, but Chicago's harbor remained in poor condition for two decades after that. The I&M Canal, proposed early in the century, was not completed until 1848.

The long lags were due to specific setbacks, including the War of 1812, the Panic of 1837, and the generally depressed conditions of business that prevailed on the western frontier much of the time. The most serious obstacle, however, was Chicago's lack of an economic role. The city produced little and exported even less. The first railroad that was chartered by Chicagoans, in 1835, was to be extended southward to Vincennes, Indiana, in the Wabash Valley. Its purpose was the importation of supplies of food that Chicago needed for its own sustenance. The impasse was broken in 1848 when the I&M Canal and the Galena and Chicago Railroad were completed. Reliable transportation stimulated the flow of goods to and from the city in the usual pattern of regional development: but, until that time, the city grew only on the faith newcomers had in what it might become.

Chapter 8
EARLY DEVELOPMENT

Sᴜᴄᴄᴇꜱꜱꜰᴜʟ ᴄᴏᴍᴘʟᴇᴛɪᴏɴ ᴏꜰ ᴛʜᴇ Eʀɪᴇ Cᴀɴᴀʟ in 1825 attracted favorable attention to canals as a means to stimulate economic development. In 1827, Congress awarded a land grant to support construction of the Illinois & Michigan Canal on an alignment that followed the Chicago and Des Plaines rivers from Chicago to Ottawa in the Illinois Valley. As with all federal land grants, sales of canal lands were to be made immediately to defray costs of construction. The initial property offerings during the summer of 1836 placed some of the land in the hands of speculators, whose sales immediately drove up the price.

The boom was cut short by the Panic of 1837, which, in turn, was prelude to a nationwide economic depression that lasted more than five years. By the end of the 1830s, town lots in Chicago had become almost unsalable. Federal payments for canal construction and harbor improvement helped the new city weather the economic crisis, but by 1841 many local merchants were bankrupt. The State of Illinois, which had inherited responsibility for building the canal, declared bankruptcy that year, and all work on the canal stopped.

Actual construction of the canal began in 1836, financed by loans and government payments. Building the canal was the first significant economic activity in Chicago, and for much of the late 1830s it was the one of the few sources of employment in the city. Contractors needed labor, and they hired some of the city's first German and Irish immigrants. Small ethnic neighborhoods soon appeared, and, as more immigrants arrived, they often settled near others with the same national origin.

According to a census taken in 1843, the Irish were concentrated along the east bank of the South Branch of the Chicago River. From that beginning, the Irish component of the population grew and spread south and west, along the route of the canal that employed so many of them. In the same census, German immigrants were found to be concentrated north of the Chicago River and east of Clark Street. The North Side German enclave and the Southwest Side Irish concentration were the nuclei for ethnic communities that expanded steadily during the second half of the nineteenth century as more immigrants arrived in the city.

When government payments were suspended and canal construction ceased in the early 1840s, the new European immigrants suffered like everyone else. The devaluation of property that accompanied economic depression resulted in a flow of money out of the city, just as it had flowed in as a result of speculative optimism. The first half of the 1840s was a period of dormancy in Chicago, and the rate of arrival of new residents was reduced. When a general recovery from the depression began during the mid-1840s, the state was determined to complete the I&M Canal as quickly and cheaply as it could. Following somewhat less ambitious plans than were made initially, the canal finally was completed and opened for traffic in the spring of 1848, barely six months before Chicago received its first railroad link to the west.

CHICAGO IN 1850

By 1850, Chicago had grown to become a city of some 30,000 inhabitants (Fig. 8.1). Nearly half of that growth had come in the previous three years, as the tempo of economic activity picked up and national attention was once more focused on developments in Illinois. The city limits followed North Avenue, Western Avenue, and 22nd Street, with a small extension north to Fullerton Avenue along the lakefront. Population growth had taken place in all parts of the originally platted area and spread outward along both the North and South branches of the Chicago River as well.

In many respects, however, Chicago was still a village in 1850. The population map based on the 1850 census suggests that people lived in all parts of the city without the apparent separation of residential, commercial, and industrial areas. The fairly low density of population, even in the original plat on the south half of section 9, indicates that a mixture of land uses prevailed, including land set aside for pasture and crops. Substantial tracts of land along the lakefront were sandy, swampy, or both and did not attract settlement. These conditions would change rapidly during the next twenty years as population grew and competition for desirable lands increased.

Chicago's "canal era"—in the sense of a phase of transportation preceding railroads—strictly speaking lasted only a matter of months. In fact, completion of both the canal and the first railroad were delayed for at least a decade because of the difficult economic times. Capital investment had to come from outside the region, and, unless the prospects for a favorable return were attractive, those with money to invest turned their attention elsewhere. Once completed through to Ottawa in the upper Illinois Valley, the I&M Canal was to play an important role in funneling wheat and corn shipments from northern Illinois to Chicago. Not until 1868 did the

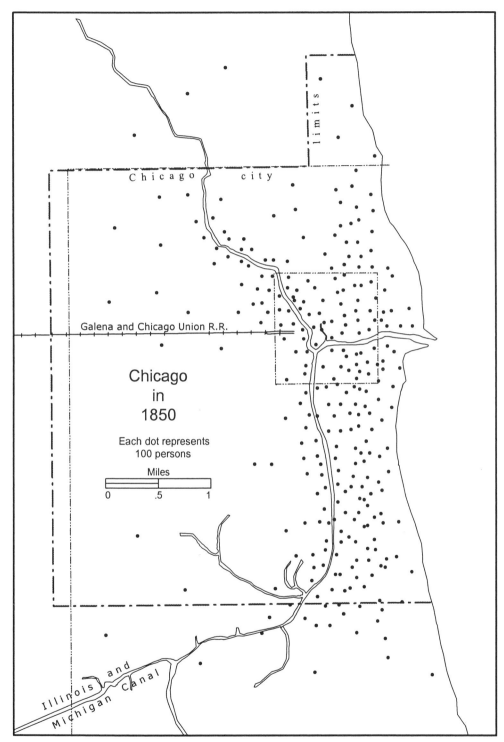

Chicago
in
1850

Each dot represents
100 persons

Miles
0   .5   1

Chicago city

limits

Galena and Chicago Union R.R.

Illinois and
Michigan Canal

Fig. 8.1

Illinois Central Railroad replace the I&M Canal as the largest source of cash corn at Chicago. The canal's traffic in other commodities lasted even longer.

STRATEGIES OF THE EARLY RAILROADS

Although Chicago's start as a railroad center was delayed by the same forces that prevented a more timely completion of the canal, a boom in railroad building took place in the 1850s. Geographically speaking, those who built railroads were divided into two groups. Eastern investors were interested in extending lines of track to the new city at the southern end of Lake Michigan, and they competed with one another to have their own line of track completed ahead of any competitor's. Eastern railroad builders had little interest in extending their tracks beyond Chicago, a circumstance that incidentally helped the city become the hub of the nation's railroad network in the years ahead.

The other group of early railroad builders was Chicago-focused. Its members staked their hopes on the city's expanding role as a lake port. The lines they projected were aimed toward the better agricultural lands north, south, and west of the city. Although the Illinois Central Railroad was chartered to build a line on a north-south alignment across the entire state, a branch of the line approached Chicago on a long diagonal, beginning in downstate Centralia, that cross-cut what would become the heart of Illinois's Corn Belt. Within the city, the Illinois Central line was extended northward along the lakefront and did not stop until it reached the south bank of the Chicago River, near the waterway's entrance into Lake Michigan (Fig. 8.2). The corn brought into Chicago by the Illinois Central was sold, stored in warehouses on the riverbank, and then sent east via Great Lakes freighters.

While most of Illinois's large corn crop was fed to meat animals on the farms where it was grown, nearly all wheat was sold off the farm. Specialized wheat growers in northern Illinois and southern Wisconsin sought efficient transportation to cities to the east where the grain was milled into flour. The Chicago and North Western Railroad was the principal carrier of wheat. Like the Illinois Central, its tracks reached as close to the lakefront as possible, terminating in a warehouse district along the north bank of the Chicago River.

The agricultural frontier had expanded west of Lake Michigan in the two decades it took for Chicago to catch up with the general tide of new settlement. Farmers in the Rock River valley were hauling grain sixty or seventy miles overland, to lake ports such as Milwaukee, Kenosha, or Chicago, at a time when no railroad offered them an alternative to the long wagon haul. Chicago's first railroad, the Galena and Chicago Union, was chartered with the idea of building northwest to the lead-mining district at Galena, Illinois. The company's tracks never reached Galena, whose mines had already passed their peak production by the middle of the nineteenth cen-

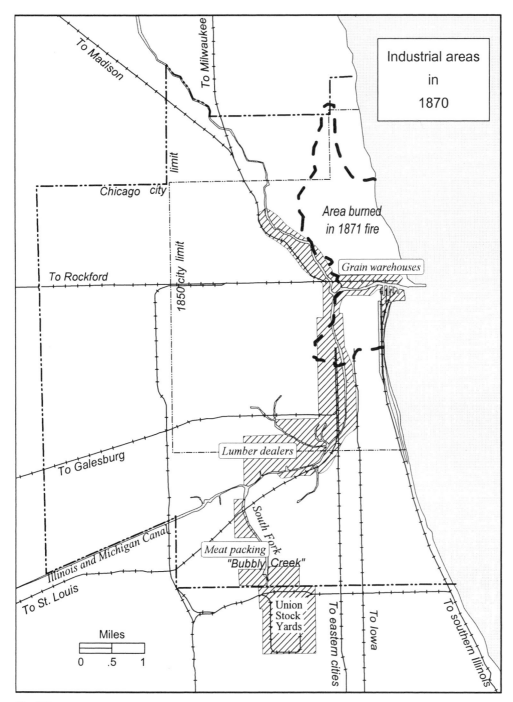

To Madison

To Milwaukee

Industrial areas
in
1870

Chicago city limit

1850 city limit

Area burned
in 1871 fire

To Rockford

Grain warehouses

To Galesburg

Lumber dealers

South Fork

Illinois and Michigan Canal

Meat packing

"Bubbly Creek"

To St. Louis

Union
Stock
Yards

To eastern cities

To Iowa

To southern Illinois

Miles

0   .5   1

Fig. 8.2

tury, but the railroad soon found a much more profitable alternative in the wheat-rich Rock River valley. Reorganized into the Chicago and North Western, the tracks reached Rockford in 1852. That same year Milwaukee entrepreneurs completed a line through southern Wisconsin's wheat fields to reach the Rock River at Janesville.

Here was an early episode of intercity corporate rivalry for trade based on railroad access that would be repeated many times in the years ahead. Milwaukee eventually became the largest wheat shipping port, but Chicago acquired perhaps an even more valuable role in 1845 when Charles M. Gray, a Chicago manufacturer of farm tools, acquired the contract to manufacture the McCormick reaper. The device had been invented by Cyrus H. McCormick to harvest wheat in the Shenandoah Valley of Virginia, and it quickly gained popularity for the labor it saved.

INDUSTRIAL GROWTH

Manufacture in Chicago of the latest in grain-harvesting machinery foreshadowed the double role the city would play in regional agricultural development. Chicago was becoming the market for the farmer's wheat and also the place where machinery needed to plant and harvest the crop was built and sold. All of these activities were accompanied by financial transactions that helped Chicago emerge as a banking center. The Chicago Board of Trade was founded in 1845. Each new economic role the city acquired enhanced its desirability as a place to concentrate new business.

With shipments to and from the city growing every year, more railroad lines radiated from Chicago that brought an increasing volume of trade. Chicago's first agricultural hinterland was the wheat-raising Rock River valley. As wheat culture began to move west with the frontier in the 1860s, Chicago declined as a primary wheat market. The expansion of corn and livestock agriculture in northern Illinois during the second half of the nineteenth century more than replaced wheat, however, because it was a system that produced both grain and meat animals.

When rivers had been the chief means of transportation, meat packers typically located their operations right on the river's bank and shipped to market down the Mississippi River or its tributaries. Chicago was at the remote end of that system, even after the I&M Canal was completed, but railroads were able to bring livestock in from every direction. The South Fork of the South Branch of the Chicago River was one site where Chicago's early meat packers congregated. The packers dumped waste and offal from the packing operations into the creek, leading to its infamous nickname, "Bubbly Creek," a result of the waste dumped there.

By 1870, Chicago was a city of nearly 300,000 people, ten times the size it had been in 1850. The two decades also saw rapid commercial growth, with new railroads and new industrial and warehouse districts appearing along the rivers and on the lakefront. Chicago's industries in this period relied heavily on primary production, especially on the output of farms and forests. Meat and livestock, grain, and lumber were three distinct lines of economic activity, and each developed its own geography of operations within the city (Fig. 8.2).

Cash grain sales drove the wheat marketing system, but some corn and other small grains were sold for cash as well. Lake transportation was essential for these activities, and thus railroad freight terminals for the receipt of grain from the country were concentrated on the riverbank where the storage warehouses also were built. Both wheat and corn also arrived at Chicago down the I&M Canal, which meant that terminal facilities for grain handling had to be accessible to the canal, the railroads, and the lakefront.

The only choice for all access to all three was to construct warehouses along the Chicago River near its entrance to Lake Michigan. The Chicago and North Western Railroad, which brought in the largest amount of wheat, dominated the north side of the river. The Illinois Central brought most of the corn and some wheat. Its operations terminated in a new development on the south side of the Chicago River.

Construction of a long pier into the lake interrupted the transport of sand that followed the southward drift of the longshore current. Sand accumulated on the north side of the pier, and the new land accidentally created became the basis for a new section of the city in later years. South of the pier, however, the sand supply was made inadequate and the shoreline began to erode. Michigan Avenue, built along what was then the shoreline, flooded when storms whipped up waves on the lake. The Illinois Central Railroad constructed a new and larger breakwater in exchange for a strip of lakefront land so that its tracks could reach north to the Chicago River. The railroad purchased what had been lands belonging to the abandoned Fort Dearborn and filled in the low-lying beach dunes for nearly 1,000 feet toward the lake. This was the beginning of a long process of infilling and breakwater construction that eventually gave the city much of its downtown lakefront land, including Grant park.

The organization of Chicago's lumber trade was in many respects the reverse of its cash grain business. Pine forests covered the interior reaches of Wisconsin and Michigan beginning about 150 miles north of Chicago. Trees were cut and floated downriver to sawmills in Lake Michigan port cities such as Green Bay, Wisconsin, and Ludington, Muskegon, and Menominee, Michigan. Sawed lumber arrived at the mouth of the Chicago River on lake vessels where it was purchased by wholesale lumber dealers (Cronon, 1991; Chapter 4). The great volume of the lumber business in the city was stimulated by the demand for building materials in newly settled areas of the mid-continent prairies and plains. These areas had railroad access to Chicago but lacked local supplies of timber.

The lumber was stored in yards that lined the banks of the South Branch on the Southwest Side of the city. There the lumber was transferred to railroad cars for distribution to cities in downstate Illinois and, eventually, to much of Iowa. Chicago dominated this trade for a time, but eventually lost it because of changes in the industry. Customers in the prairie regions to the west also had access to lumber produced in northwestern Wisconsin and Minnesota because of commerce on the Mississippi River. Exploitation of the northern forests began along the river's tributary to Lake Michigan, but, as the cut proceeded, inland areas accessible only to railroads became the largest producers. Chicago lost its advantage in the lumber trade because a dwindling fraction of the total cut moved aboard lake vessels.

Water transportation played only a minor role in meat packing, which was Chicago's third major industry developed in the second half of the nineteenth century. Live animals could be shipped via the canal, but most livestock was walked to market in the early years and was transported by rail after the early 1860s. Livestock, mostly cattle and hogs, were the most important traffic coming to the city by rail and then shipped out via the same means. Chicago's first stock yards were built south of the city along Lake Michigan in 1856, but in a location that was inaccessible to most of the entering railroads. By 1864, there were half a dozen stock yards in Chicago which operated in conjunction with meat packers who typically located as near one of the yards as possible.

The Chicago Union Stock Yards was organized in 1865. Built on 320 acres of poorly drained land along the southern city limits, the new facility was accessible to all of the railroads and had space for meat packing plants as well (Wade, 1987). More than a million and a half meat animals were received at Union Stock Yards during its first year of operation, and in 1880 eight million head arrived. By 1890, the total reached thirteen million animals a year. The stock yards and packing plants employed thousands of workers, many of whom were newly arrived immigrants, especially from Eastern Europe. They settled in neighborhoods near the stock yards and thereby created a new industry-linked pattern of ethnicity.

Philip D. Armour and Gustavus F. Swift were foremost among the entrepreneurs who organized the meat business in Chicago. They also established branch plants in distant cities. About two-thirds of the livestock arriving in Chicago were reloaded for shipment elsewhere, mainly to Eastern cities, before refrigeration was a possibility. Armour and Swift pioneered the technology for cooling and ventilating a railroad car to make it possible to ship dressed meat to distant markets. By 1890, Chicago accounted for nearly half of the urban wholesale meat business in the United States.

Almost half of Chicago's 1870 inhabitants had been born in foreign countries. Foreign-born residents had a substantial presence in all of the city's twenty wards in 1870, and they constituted a majority in the wards along the west banks of the Chicago River (Fig. 8.3). Neighborhoods along the river had a mixture of residential

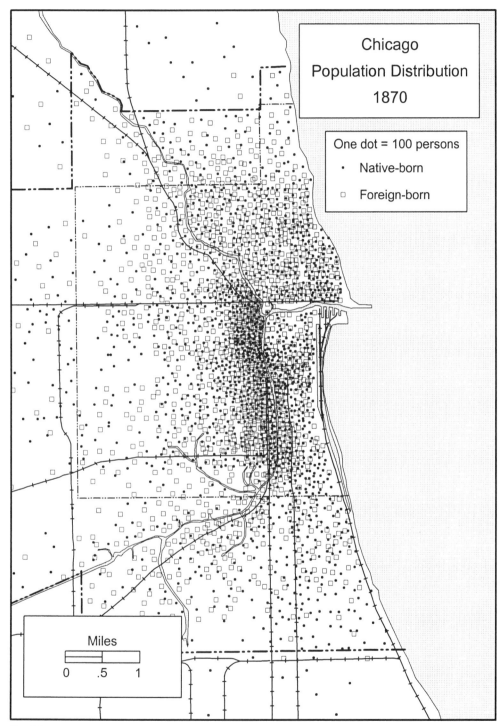

Fig. 8.3

and industrial/commercial properties. Many people no doubt were employed at or near the places they lived. Such patterns were characteristic of many cities prior to the development of public transportation and improved roads.

POPULATION IN 1870

The great influx of immigrants between 1850 and 1870 was directed toward the industrial areas, although many foreign-born residents could be counted in the outlying portions of the city where they likely were engaged in farming or at least in the production of food for their families. African-Americans accounted for little more than one percent of the city's population in 1870. There were no African-American neighborhoods as such, although most black residents lived in the original city plat, in what is now the Loop.

Meat, grain, lumber, and railroads were Chicago's first large industries. Along with these businesses which brought trade to the city were others, such as wholesaling, retailing, and commission-order sales, activities that made use of the same means of transportation and took advantage of the same financial linkages. The consumer durable goods purchased through Chicago dealers and jobbers often were manufactured in the city as well. The growth of all of these industries was not uniform from year to year, but rather followed the same booms and busts felt by the nation as a whole.

Local events also could interrupt the cycle. On the evening of October 8, 1871, a fire broke out in the Irish neighborhood near the Chicago River (resulting from a lantern kicked over by a cow in Mrs. Catherine O'Leary's barn, as the local legend has it). Over the next day, the fire raged northward, destroying much of the downtown area and spreading north to the city limits. Some 300 people lost their lives and 90,000 were left homeless.

The "Great Chicago Fire" had devastating consequences, but it also provided the opportunity for new growth and the chance to correct some of the excesses of the more than twenty years of rapid growth that preceded it. Having outgrown its simple small-town origins, it was time for Chicago to rebuild along new lines. For nearly forty years after the fire, Chicagoans turned attention toward redesigning their city in ways that were appropriate for a large metropolis.

Chapter 9
SOUTHWARD EXPANSION

Chicago's population tripled and its area more than doubled between 1870 and 1890. The city limits were pushed southward nearly 100 blocks, from 39th Street to the Little Calumet River (Fig. 9.1). On the north, Chicago expanded some forty blocks, from Fullerton to Devon avenues. Much less territory was annexed on the West Side during this period because of the appearance of new suburbs, although some of them were later absorbed by the city. Chicago acquired its elongated shape during these two decades, and, apart from small additions of territory on all three sides, it has remained roughly the same size and shape ever since.

The area annexed on the south (some ninety square miles) was the largest expansion the city would make in a comparable time period. It also apparently exhausted Chicagoans' ambitions for growth in that direction. Some growth was diverted southward after the Great Fire of 1871 destroyed much of the North Side. The South Side was, if anything, flatter than the north, and there were few obstacles to expansion. Nor were there many new settlements to the south that would have blocked the city's annexation plans.

The main reason for Chicago's southward expansion was industrial growth. Clusters of heavy industries began to appear along the southern margins of the urban area in the 1860s. Like most cities, Chicago's early focus had been on its downtown area, where commercial and industrial land uses mingled with a considerable amount of residential housing. Riverfront access initially dictated the downtown industrial concentration. When railroads came and Great Lakes navigation was improved, river, lake, and rail transportation all required access to one another.

While the commercial core of any city is typically found in its downtown area, there is no need for heavy industries to congregate in the heart of the city. Chicago's industrial zones began to expand southward, along the I&M Canal and up the North

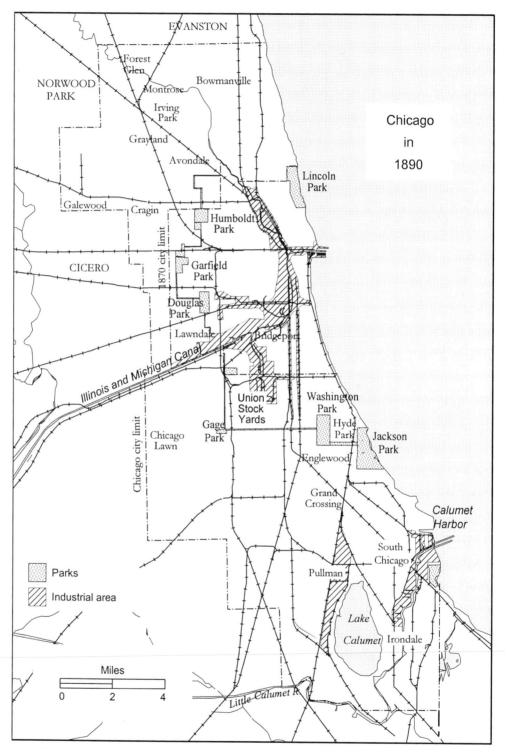

Chicago
in
1890

EVANSTON

Forest Glen

NORWOOD PARK

Montrose

Bowmanville

Irving Park

Grayland

Avondale

Lincoln Park

Galewood

Cragin

Humboldt Park

1870 city limit

CICERO

Garfield Park

Douglas Park

Lawndale

Bridgeport

Illinois and Michigan Canal

Chicago city limit

Chicago Lawn

Gage Park

Union Stock Yards

Washington Park

Hyde Park

Jackson Park

Englewood

Grand Crossing

Calumet Harbor

South Chicago

Pullman

Lake Calumet

Irondale

Parks

Industrial area

Miles

0   2   4

Little Calumet R.

Fig. 9.1

Branch of the Chicago River, in the 1870s. Creation of Union Stock Yards in 1865 attracted the livestock and meat-packing industries to the new industrial area south of 39th Street. Grain warehouses and other shipping facilities remained in the downtown lakefront area for a time, although they, too, eventually moved south.

CALUMET HARBOR

The initial movement of industries to the southern fringes of the urban area began in 1869 when the federal government appropriated money to improve the mouth of the Calumet River to permit Great Lakes ships to dock there. The new site lay some twelve miles down the lakefront from the mouth of the Chicago River, but there were no intervening locations suitable for a harbor. The shallow near-shore depth of Lake Michigan and the lack of any river mouths between the Chicago and the Calumet meant that the intervening stretch of land was unsuitable for harbor development. That, in turn, preserved Chicago's South Side shoreline for other uses.

Waterway improvements to connect Calumet Harbor with the Illinois River would not come until the 1920s, but there were some immediate benefits from the new lakefront location to the south. Grain warehouses could be built in the Calumet district because it had access to the entering railroads. Railroad yards lingered in downtown Chicago for years thereafter, but bulk commodity shipping eventually disappeared from downtown after Calumet Harbor was built. More important than the removal of old industries from the city center was the creation of new types of industry that needed space as well as access. By 1874, iron furnaces and steel mills had appeared around Calumet Harbor. Ships loaded with the coal, iron ore, and limestone necessary to operate an iron and steel industry were able to shuttle between sources of those raw materials around the Great Lakes region and unload their cargoes adjacent to the new mills.

From the 1870s through the early decades of the twentieth century, the new steel industries expanded from Chicago to the adjacent shoreline of northwest Indiana. The whole complex of mills and factories, a zone eventually known as the "Calumet region," became one of the nation's largest producers of iron and steel. But the sandy, marshy shoreline of northwest Indiana was no more able to accommodate Great Lakes shipping than was the Illinois side, and the lake's margins were reworked numerous times—even into the 1970s—for deep-water harbor facilities.

PULLMAN

In 1880, railroad-car manufacturer George M. Pullman relocated his local factories to a 3,500-acre tract of land that was well south of Chicago's city limits. The Pullman Company built factories and also constructed houses and apartments that were rented to company employees. Pullman had its own stores, hotels, and schools. Food for the community was produced on company-owned farms. By 1884, the Pullman community had a population of 8,000 (Fig. 9.1).

As an experiment in social betterment, Pullman attracted favorable attention for a time because, as a "company town," it offered an alternative to some negative aspects of industrial growth. The community was annexed by the City of Chicago in 1890. During the nationwide economic depression of 1893, the Pullman Company reduced the wages paid to its employees but kept constant the rents it charged to live in company-owned housing. The fledgling American Railway Union, under the leadership of Eugene Debs, called a strike against Pullman, which brought nationwide attention and prompted President Grover Cleveland to send federal troops. Although Debs lost the fight and was jailed for a time, he and his cause gained great popularity. In 1898, the Illinois Supreme Court effectively ended the experiment by ordering that Pullman sell the town, much of which was purchased by the company's employees.

Pullman was an outlier of heavy industry on the far South Side of the city, but it eventually served more as a model to be avoided rather than as one to be copied. It was built on the marshy fringes of Lake Calumet, several miles from Lake Michigan, and thus was inaccessible in terms of receiving and shipping heavy bulk commodities. Industrial sites fringing Lake Michigan were more attractive, and this is where all of the major steel manufacturers erected new mills after 1900.

The merger of the Andrew Carnegie and J. P. Morgan interests in 1901 created the United States Steel Corporation. One of U.S. Steel's first major projects was construction of an integrated iron- and steel-making complex on what had been acres of dune lands in Lake County, Indiana. In 1907, the company's real estate subsidiary laid out its new town of Gary, named after the company chairman, Elbert H. Gary. The new city of Gary was a company town, but the real estate was not owned by U.S. Steel, the company deliberately avoiding the paternalism of George M. Pullman.

PARKS, BOULEVARDS, AND THE WORLD'S FAIR

With heavy industries growing on the southern fringes of the city and an established industrial zone along both the North and South branches of the Chicago River, Chicagoans began to plan more deliberately for the future of their sprawling metropolis. The city's commercial growth had been so rapid that little attention had been given to the creation of parks, boulevards, and other amenities. In 1869, three independent park commissions were established—one each for the North, West, and South sides of the city. Landscape architect Frederick Law Olmsted and his colleague, architect Calvert Vaux, were hired to design a park system.

With the assistance of the landscape architect, Jens Jensen, Olmsted and Vaux created Humboldt, Garfield, and Douglas parks on the city's West Side. Lincoln Park, slightly older, was to be the North Side's park. Olmsted and Vaux also created Jackson and Washington parks on the South Side, although both of them were beyond the city's limits at that time. The most striking feature of the plan was a series of boulevards linking all of the parks in a great semicircle roughly three to four miles away from the city center (Fig. 9.1).

Humboldt, Garfield, and Douglas parks were bounded by city streets and fit into the urban grid. The North Side's Lincoln Park, which in later years expanded northward along the lakefront, featured lagoons somewhat along the lines of New York City's Central Park, which also had been created by Olmsted and Vaux. Jackson and Washington parks on the South Side were open, marshy land with a covering of dune sand at the time of their designation as parks.

In 1890, the federal government chose Chicago as the site for a great fair, the World's Columbian Exposition, to commemorate the 400th anniversary of the coming of Christopher Columbus. Olmsted was hired to be the fair's chief architect and he recommended Jackson Park, still largely undeveloped, as the site. The fair's chief of construction, Daniel H. Burnham, commissioned sculptors Augustus Saint-Gaudens, Daniel French, and Lorado Taft to create statuary. The architects, artists, and designers brought together by Burnham created a memorable new city of largely artificial facades—the "White City." To many, the great fair of 1893 was the high point in Chicago's history.

Olmsted linked the refashioned Jackson Park with Washington Park via a long boulevard, which he named the Midway Plaisance, incorporating all of it into the fair's design. The Midway, which later became a major thoroughfare adjoining the University of Chicago's campus, was only one of nearly a dozen such boulevard segments designed to link the city's nineteenth-century parks. The Columbian Exposition brought a population boom to nearby neighborhoods, especially Hyde Park. The fair itself was dismantled in the mid-1890s, although the Palace of Fine Arts was later rescued and rebuilt as the Museum of Science and Industry.

Chicago's city limits expanded dramatically during the two decades following the Great Fire of 1871, but the city's population remained concentrated within the boundaries that existed in 1870. Population expanded rather evenly in all directions, away from the city center, with the most rapid growth taking place directly north and directly south of downtown. Newly annexed neighborhoods on the Northwest Side of the city were only thinly populated in 1890, while other new areas, especially around the South Side industrial zones of Pullman and Calumet Harbor, were growing more rapidly.

POPULATION ORIGINS

Of the more than one million inhabitants the city could count in 1890, slightly more than two-fifths were of foreign birth. The overall patterns of native-born and foreign-born residences were similar (Figs. 9.2 and 9.3). Native-born Chicagoans were more scattered over the city's total area, yet they remained strongly concentrated in a

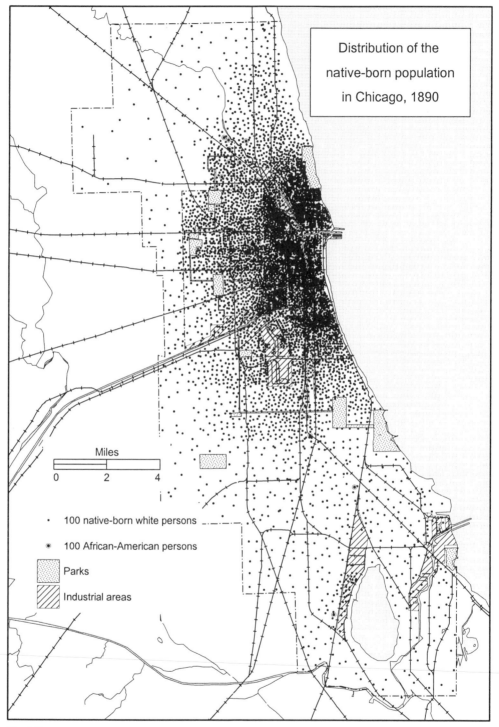

Distribution of the
native-born population
in Chicago, 1890

Miles

0   2   4

·   100 native-born white persons

⊙   100 African-American persons

Parks

Industrial areas

Fig. 9.2

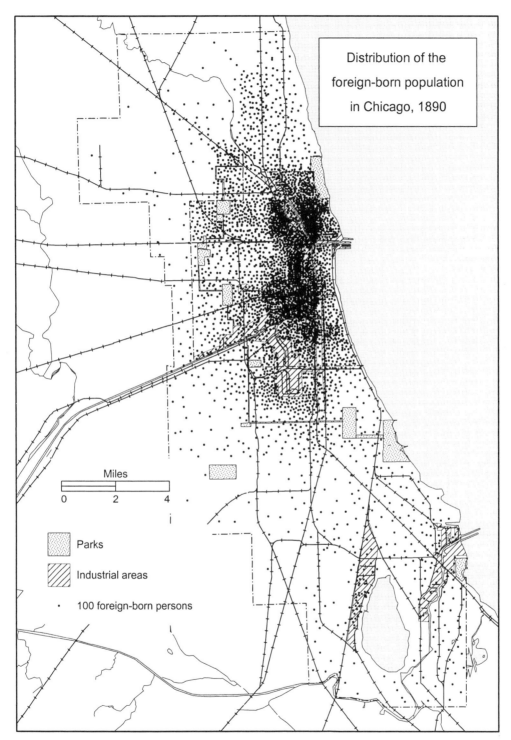

Distribution of the
foreign-born population
in Chicago, 1890

Miles

0   2   4

Parks

Industrial areas

· 100 foreign-born persons

Fig. 9.3

zone within three miles of the city's center. Although commercial, industrial, and residential neighborhoods were evolving in separate locations by this time, older housing remained in the city's center. Long-time Chicago residents who were moving outward by 1890 passed their older housing along to new arrivals, starting a process that would continue for years to come.

Expansion of the city on this scale was made possible by the growing network of street-car and commuter railroad lines. The early horse-drawn trolleys were replaced with electric street-cars. Improved transportation extended the zone of daily commuting to a radius of roughly three or four miles from downtown. New residential suburbs appeared five to ten miles from downtown along the major railroad lines, beyond the zone of street-car commuting. Suburban expansion was closely linked to employment growth in the city's center, which increasingly meant white-collar residents would live in the more distant residential zones.

The residential pattern of foreign-born Chicagoans reflects some of the same centripetal tendencies, but the foreign-born were more likely to be employed in blue-collar occupations and were more likely to live near the industrial areas where they worked (Fig. 9.3). Industrial growth in the stock yards, Pullman, and Calumet Harbor areas drew increasing numbers of the foreign-born. More than half of the foreign-born in 1890 worked in the congested zone of light industries, wharves, yards, piers, and warehouses that crowded both sides of the Chicago River.

Chicago was divided into thirty-four wards in 1890, but only six of them had more foreign-born than native-born residents. One such area of foreign-birth dominance was the new industrial area near Calumet Harbor. The other five wards were in older downtown areas, especially the west side of the South Branch and both sides of the North Branch of the Chicago River, near the confluence of the two streams. This was the old city, which had become a crowded neighborhood of substandard housing by 1890.

The Census of 1890 does not list country of birth by city wards for the foreign-born, but the pattern is known because, once established, ethnic neighborhoods tend to persist except, perhaps, in the face of overwhelming challenges. The Irish, who made up Chicago's first ethnic enclave in the Bridgeport neighborhood, had become scattered over more than a dozen areas of the city, from the far South Side to the North Side. Germans, who were numerically the largest of the non-English speaking immigrants, remained concentrated on the North Side and were overwhelmingly the most important population subgroup east of the North Branch of the Chicago River. By 1890, Poles had established their own neighborhoods, in fairly tight, wedge-shaped sectors on the Northwest, West, and Southwest sides of the city, from whence they would expand greatly in the decades to come.

The year 1890 also marks a transition in the foreign-born population. From that time onward, an increasing share of the new arrivals would come from Eastern and Southern Europe and would, in turn, displace some longer-established foreign-born residents from neighborhoods west of the Chicago River. Each population group established its own residential pattern. Some groups concentrated in tight enclaves while others dispersed over a broad area.

Fewer than 15,000 African-Americans lived in Chicago in 1890. They made up slightly more than one percent of the population, the same fraction as in 1870. The great northward migration of African-Americans from the Cotton South had yet to begin. Nearly two-thirds of the city's African-Americans lived in the first, second, and third wards, from the Chicago River south to 33rd Street. They were also scattered over nearly all of the other wards of the city, roughly in proportion to the total population. It is difficult to discern an actual African-American neighborhood, partly because of the small numbers, but also because black people lived and were employed in nearly all parts of the city. The rigid segregation that eventually would typify African-American residential patterns in Chicago had scarcely begun in 1890.

Nor were any of the foreign ethnic groups "segregated" in any meaningful sense. The map of foreign-born Chicagoans is composed of many nationalities, each with a distinctive pattern, but there were no areas where only immigrants lived nor were there any areas totally without them. As was true in 1870, fewer foreign-born residents lived east of the Chicago River and fewer lived in outlying areas, but no ward in the city had a foreign-born component less than one-third of the total population.

The 1890s were years of economic stagnation in many parts of the United States, but Chicago continued to boom, adding nearly 600,000 new residents during the decade. The city contained more than ninety-two percent of Cook County's population in 1900, a percentage that had been growing for several decades as the city annexed more and more territory. By 1950, Chicago's share of Cook County's population would fall to around eighty percent, as the suburban growth boom got underway. But in 1900 the city was still concentrating population, even in its inner-city neighborhoods.

Of the suburbs, Evanston, with nearly 20,000 inhabitants in 1900, was the largest (Fig. 9.4). Unincorporated Cicero was smaller by a few thousand. Blue Island's 6,000 inhabitants made it the largest suburb on the South Side; it was followed in size by Chicago Heights, which had a population of slightly more than 5,000.

Population had only begun to spread outward along the radiating commuter railroad lines by 1900. Linear patterns are evident to the north, northwest, and west of the city, although nearly all of those future suburbs had populations under 1,000 and none could count more than 2,000 inhabitants in 1900. Large areas within Chicago

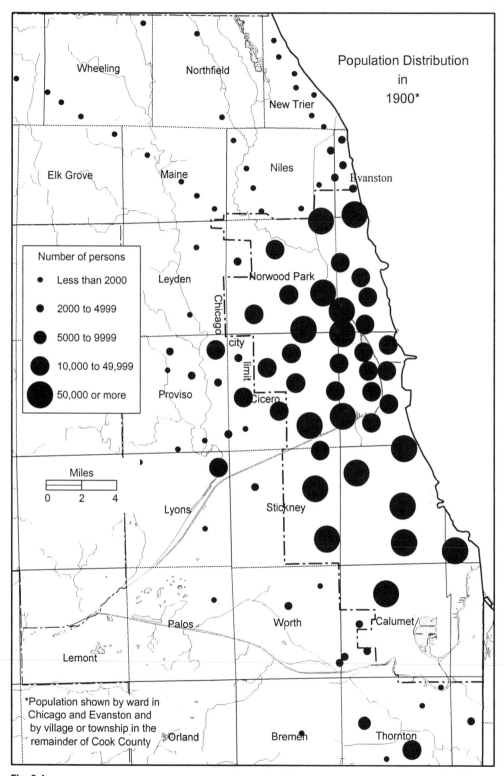

Fig. 9.4

had only scattered settlement. Until the city began to run out of room for new housing developments, expansion beyond the city limits was not needed.

The closing decades of the nineteenth century witnessed a net shift in Chicago's population toward the south. In 1900, forty-two percent of the city's population resided north of the line of Lake Street and the Chicago River, with fifty-seven percent south of that line. Roosevelt Road (12th Street), on the South Side, roughly approximated the break-even line, with half the city's population to the north and half to the south.

Reasons for the southward shift included the growth associated with the World's Fair, industrial growth along the southern fringes of the city, and the simple statistical fact that more territory and population had been added on the South Side than on the North in the preceding decades. The apparent southward shift interested city planners who were about to launch a new phase in Chicago's development that would refocus the city slightly southward, following what seemed to be the future trend.

Part IV

THE GROWING CITY

Chapter 10
MODELS AND PLANS

IF CHICAGO IS REGARDED AS A TYPICAL CITY in the sense that its geography reflects certain urban processes that might be expected to operate anywhere, then attention also must be given to the role of deliberate planning. Did Chicago's patterns simply emerge, or were they guided by planners? Chicago began with a simple grid of city blocks that might have been found in any early nineteenth-century American city. As the city grew, influential residents began to take an interest in reshaping their surroundings to accommodate the growth that had taken place or was yet to come.

If the concentric rings, sectors, and wedges that eventually became part of the Chicago model were somehow already there, waiting to emerge, they would not have needed to be planned. Beginning with the rebuilding process that followed the fire of 1871, however, Chicago's landscape increasingly reflected the influence of deliberate planning. The early decades of the twentieth century were a time of keen interest in city design. The plans launched at that time would guide Chicago's development for many decades to come.

In 1909, Daniel H. Burnham and Edward H. Bennett published their *Plan of Chicago*, a beautifully illustrated volume that offered a comprehensive approach to reshaping the urban environment (Burnham and Bennett, 1909). Popularly known as the "Burnham plan," the document was adopted by the City of Chicago as its official plan in 1910. The book itself was widely read and enthusiastically received by Chicagoans. It endured for many years as an outstanding example of urban planning that combined natural features, technological advances, and an ambitious architectural scheme into a single product. Although only a portion of what Burnham proposed actually was built, the ideas first put forth in the volume would have a lingering influence. Nearly a century later, the Burnham plan serves as a benchmark against which developments of the twentieth century can be measured (Fig. 10.1).

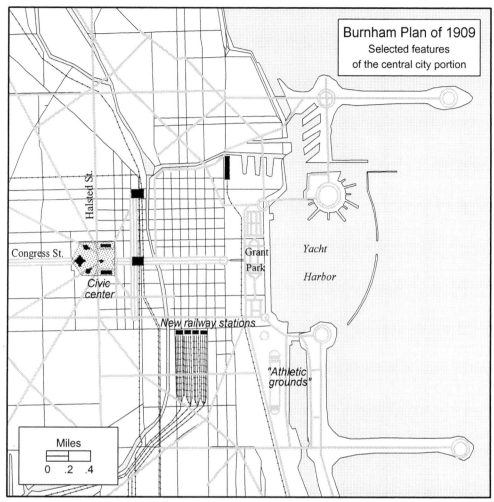

Fig. 10.1

DOWNTOWN CHICAGO

In some respects, Burnham's "City Beautiful" plan was a projection onto the entire city of what he had achieved in planning the World's Fair of 1893. The book was filled with architectural drawings that imitated the grand designs of classical antiquity. Following the ideas of Renaissance urban planning, space was a formative element of the plan and not just the interstices between buildings. Burnham proposed broad avenues and boulevards, some radiating from circular foci. He favored diagonal thoroughfares as a means of decongesting traffic that otherwise would have to move in rectilinear fashion across the traditional grid. The boldest feature of the plan was a new Civic Center, to be constructed at the intersection of Congress and Halsted streets, that would become the focal point of the city. In Burnham's words, the new civic center would be a "monument to the spirit of civic unity."

Unity was a hallmark of Burnham's approach. He regarded the city as a "complete organism in which all its functions are related one to another in such a manner that it will become a unit." Classical styles, incorporating their own symmetries and other repetitive motifs, were well suited to the long perspectives that Burnham's streets and boulevards opened to public view. The central focus on the proposed civic center was supported by half a dozen radiating thoroughfares that led to secondary foci.

Burnham's Chicago plan bears an obvious resemblance to the layout of Washington, D.C., with its radiating boulevards focused on the national capitol. In fact, Burnham chaired the 1901 Park Commission of the District of Columbia that updated Major Pierre L'Enfant's 1791 plan for Washington. Burnham's Washington group included other architects and landscape architects who had been involved in planning for the World's Fair of 1893. As part of their work, the group visited Paris and Versailles, the acknowledged French precedents for L'Enfant's original plan.

Nothing even remotely approaching Burnham's grandiose civic center ever was built at the intersection of Congress and Halsted, although the site was later reclaimed for public purposes as the University of Illinois at Chicago's campus. Nor was the elaborate "yacht harbor," with its many slips and moorings for pleasure boats constructed east of Grant Park. But Congress Street, the main thoroughfare west from downtown Chicago, was upgraded and eventually became the Eisenhower Expressway. The double-deck highway along the Chicago River was built and is now Wacker Drive. Burnham's "athletic grounds" became Soldier Field. His plan made space for the major museums, including the Field Museum, Adler Planetarium, and Shedd Aquarium, all of which occupy monumental buildings done in the classical style. Of the two long projections of landfill into Lake Michigan, the southern one was not built, but its northern counterpart became Navy Pier, which by the end of the twentieth century had evolved a mix of business and recreational activities of the sort Burnham envisioned. Grant Park's reconstruction, with its central Buckingham Fountain, similarly fits into the Burnham Plan.

Burnham was concerned with planning a livable city that would comfortably accommodate several million people. The slicing thoroughfares were a deliberate attempt to reverse the trend toward increased congestion by speeding up the flow of traffic and creating a sense of openness in the urban grid. The alternative would have been unthinkable at the time: deconcentrate and disperse the city by removing functions and activities from the center and relocating them to the fringe. This latter approach would come in time, largely after World War II; but, before the automobile was a factor, such a dispersion could not have been achieved. The hierarchy, axiality, balance, and long perspectives Burnham proposed were a way of reducing congestion without sacrificing the city's essential focus.

After buildings and streets—which any city had to have—the next most important use of land in central Chicago was railroad tracks, yards, and stations. Despite the existence of a "Union Station," Chicago had six separate stations in its downtown area for much of the late nineteenth and twentieth centuries. Burnham proposed removal of some of those stations with a consolidation of facilities in a massive new terminal to be built on 12th Street (Roosevelt Road). Trackage associated with the other stations (north of Grant Park and along the Chicago River) was to be reduced as well. He also proposed a new terminal near the new civic center, at Congress Street. The in-facing arrangement of railroad stations would have provided rapid access to the city's commercial center.

Even the lingering notion that Great Lakes ships had to be accommodated in downtown Chicago was built into the plan in the form of new berths near the Chicago River. Burnham could not have foreseen that access to railroad stations or to lake boat piers would soon matter much less than it did in the 1900s. Nor could he have imagined that the location of an airport in the years to come would have the effect of tugging the entire economy of the city in a new direction. As with all plans, designs have a much longer life than does the technology on which they are based.

Foremost among the centralizing features accommodated (but not first proposed) in the Burnham plan was the downtown loop of elevated railroad tracks built over Lake, Wabash, Van Buren, and Wells streets. "The Loop" was literally that—a means for turning elevated trains when they arrived downtown so that they would then be headed in the direction from which they had come. The technology to build and operate such a system of electrically powered rapid transit trains was available by the 1890s. By 1910, rapid transit lines converged on the Loop from Jackson Park, the Stockyards district, and Englewood on the south; from Oak Park on the west; and from Wilson Avenue on the north. This system, which was substantially augmented with subway construction beginning in the 1940s, has long exerted a powerful centralizing effect on the city that has created an equality of access from all directions. Its original geometry was imposed upon Burnham's plan, but the rapid transit system's simple structure and single focus fit comfortably into their overall plan.

Burnham regarded the Lake Michigan shoreline as the city's most important asset, and he fought to reclaim it for the general public. His plans envisioned a lakefront park extending the entire length of Chicago's shoreline—and beyond that even to Wilmette. Lincoln Park eventually was extended northward nearly to the city limits, Grant Park was reshaped and broadened, and the pieces of lakefront park to the south were assembled into nearly an unbroken strip, all much in the fashion that Burnham recommended.

Conflict arose as to whether downtown lakefront land should be kept largely vacant for parks or partly set aside for public institutions. The latter alternative was followed, and most of the building took place in a short period of time. The Field Museum of Natural History and the Art Institute of Chicago were created in or next to Grant Park in the mid-1890s. The Museum of Science and Industry, endowed by Julius Rosenwald, president of Sears Roebuck and Company, was created out of the Palace of Fine Arts that was built in Jackson Park for the World's Fair of 1893. In 1925, John G. Shedd, president of the Field department store, donated money to build the Shedd Aquarium. Three years later, Max Adler, an executive of Sears Roebuck, donated money to begin the Adler Planetarium. The location of these institutions either were included or anticipated in the Burnham plan.

Daniel Burnham and landscape architect Jens Jensen also were influential in establishing a more rural park system that became the Cook County Forest Preserve District. In 1909, the Illinois legislature approved the creation of forest-preserve districts as land that would be set aside as parks for the people and forever remain undeveloped. Most of the land now included in the preserves lies along the Des Plaines River or the North Branch of the Chicago River, although small tracts have been set aside in scattered areas throughout Cook County. These areas, intended to be as natural as possible, were championed by Burnham because of the green space they provided near the city.

INDUSTRIAL ZONES

The comprehensive approach that Burnham and Bennett took in city and regional planning included suggestions for reorganizing Chicago's pattern of industrial land use (Fig. 10.2). The original downtown and lakefront industrial areas were fading by 1900. They had been replaced by new areas of manufacturing, wholesaling, and warehousing activity along the freight railroad lines. Burnham proposed a concentration and partial relocation of these activities into what he termed "ultimate locations" for industry.

One corridor followed the tracks of the Chicago, Milwaukee & St. Paul Railroad directly west from downtown to Cicero Avenue (4800 West). From there the corridor was directed southward along the Belt Railway of Chicago's tracks to the new Chicago Sanitary and Ship Canal, which was completed in 1900. The other two sides of the zone were formed by the canal and the North Branch of the Chicago River.

The largest of Burnham's "ultimate zones" for industry was a new tract adjacent to the Southwest Side city limits. Known as the Clearing Industrial District, the area was designed for access to railroad freight transportation. Clearing, which eventually saw the construction of several large yards for the interchange of traffic between eastern and western railroads, has remained an almost exclusively industrial enclave, long since surrounded by residential areas.

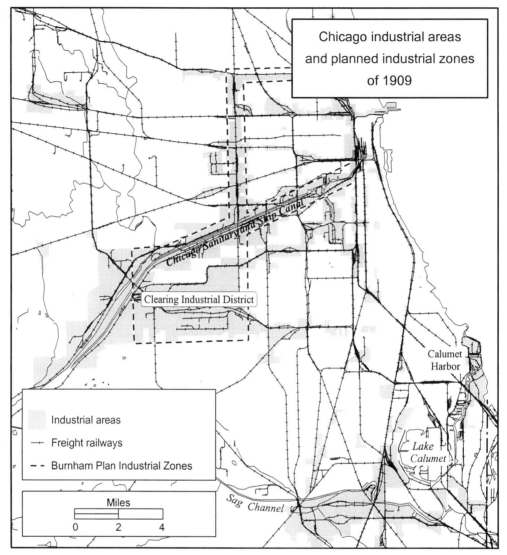

Fig. 10.2

Large-scale removal of industrial activities from the city was not a goal of the Burnham plan because that would separate homes from workplaces in an artificial manner. The creation of a new industrial zone, with possibilities for new residential neighborhoods in the surrounding undeveloped lands, was more characteristic of Burnham's approach.

REGIONAL PLANNING

The Burnham plan was one of the first proposals for restructuring any city that also incorporated aspects of regional planning for a much larger area. Burnham extended the scope of his plan fifty miles outward from downtown Chicago and thereby created one of the first maps displaying a "Chicago region" of a sort that would become familiar in the years ahead (Fig. 10.3).

The radiating diagonal streets of downtown Chicago had their counterparts in the outer region. Although Burnham stressed that most of the roads he included in his plan already existed, he added new diagonals and circumferential highways to complete the picture. Four circumferential highways, spaced about ten miles apart, were to encircle the city. The innermost circle would have required substantial new road construction. Moving farther out, he pieced together the circumferential highways largely from segments of the existing road network. He also deleted many north-south or east-west segments from his map to give the sense of a concentric focus.

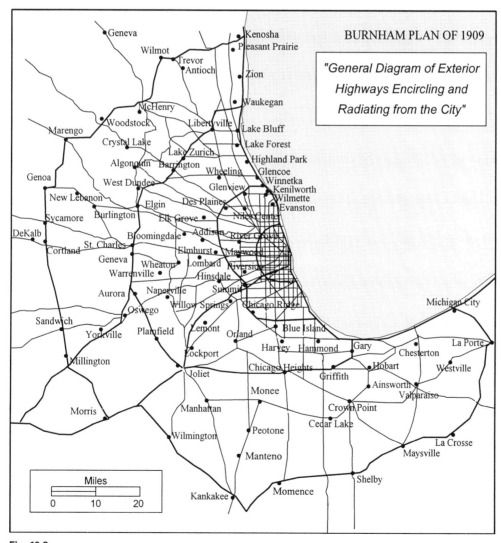

Fig. 10.3

All of the actual communities that Burnham included in his *Figure XL* are shown in Fig. 10.3 (Burnham and Bennett, 1909, Plate XL). While some of the communities were substantial settlements, others were little more than crossroads hamlets and probably were chosen for their location on either the circumferential or radial routes. Many existing communities were left off the map, perhaps because their locations did not suggest the importance of the spoke-and-wheel "Garden City" idea Burnham wished to present. Limits of the region included Kenosha, Lake Geneva, Woodstock, De Kalb, Morris, Kankakee, La Porte, and Michigan City—a fair approximation of the outer limit of the metropolitan area, whether in Burnham's time or today.

Although the regional map seems to foreshadow many actual developments that would come later, including circumferential expressways, radiating freeways, and a sprawling network of satellite communities, closer inspection reveals that the resemblance is rather superficial. As Burnham saw it in the early years of the twentieth century, one of the principal uses of the outlying highways would be recreational. People could drive (in horse-drawn or horseless carriages) along the meandering Kankakee River, for example, or alongside the Indiana Dunes. Other Burnham highways followed great arcs across the fertile farmlands of northeastern Illinois, aimed at no particular destination.

Burnham viewed the outer region as one of agricultural production that supported the city, and he shaped it with a park-like appearance to emphasize what city dwellers might appreciate visiting. But few of the actual suburbs and expressways of later years are to be found on Burnham's map. Missing is the intensive zone of bedroom communities, evident by 1950, in the ring between Burnham's first and third encircling highways. No awareness that these middle zones would soon be occupied by thousands of daily suburban commuters is evident in the plan.

One of Burnham and Bennett's primary goals was to reshape the city for the sake of its inhabitants. Open space, green space, and ease of access all contributed to the general health and well-being of those who lived in the metropolis. Extension of the plan fifty miles into the surrounding countryside was undertaken with the same goal in mind. Even the axial symmetries and central focus employed in the Burnham plan for downtown Chicago have their counterparts in the countryside.

Given the widespread popularity of the Burnham plan and the many aspects of it that actually were implemented, to what extent was it a blueprint for Chicago's later geography? One of the reasons Burnham's ideas enjoyed a favorable reception was that his plans incorporated a good deal of what was already underway in Chicago. Some proposals called for tearing down and starting over, but for the most part Burnham avoided radical changes in favor of more subtle redirections of building and expansion plans already in progress. The Clearing industrial district was one such example. The area had been considered for some time as the future site of a large railroad yard for the interchange of freight traffic. Burnham's proposal broadened the idea to include a new location for manufacturing industries as well.

The many symmetries and repetitions of self-similar geometries Burnham favored were comparatively easy to impose on Chicago. From the city's earliest start, with a small grid of city blocks that straddled the confluence of the two branches of the Chicago River, there had been a symmetric natural division into North, West, and South sides. When park commissions were established for those three sides of the city in the 1870s, and Olmsted and Vaux laid out their scheme of parks and boulevards, those, too, followed a symmetric structure. Burnham incorporated this history, of course, and he both enhanced it within the city and extended it beyond the city limits into the countryside. The Burnham plan helped keep Chicago's overall structure and layout on the course that had been evident from the start.

Chapter 11
POPULATING AND REPOPULATING THE CITY

CHICAGO'S POPULATION GREW by an average of 500,000 persons per decade between 1880 and 1930. As in most large cities, growth virtually ceased during the Great Depression of the 1930s, then rebounded during the 1940s. But the great booms that had drawn hundreds of thousands of migrants to the city in earlier times had come to an end by 1950. Most growth after that time was in the suburbs, while the central city entered a long period of either slow growth or population loss. At the middle of the twentieth century, Chicago's population of 3.62 million was a diverse mixture of native- and foreign-born people that included nearly half a million African-Americans, many of whom had migrated to the city in the preceding twenty years.

RESIDENTIAL GROWTH, 1900-1920

Growth of the city around its initial core was remarkably balanced in all directions (Fig. 11.1). Settlement spread more rapidly to the north and south than toward the west, but there was no directional bias in the overall pattern of growth. The initial corridors of expansion lay along the North and South branches of the Chicago River, a tendency that was further enhanced by the construction of railroad lines after 1850. By 1870, scattered outliers of urban population had appeared at some distance from the central city, and these, in turn, became secondary foci around which settlement expanded in later years.

The great annexations of territory that took place between 1880 and 1900 stimulated new population booms that filled in the areas between radiating transportation routes. A thin scatter of residential neighborhoods spread outward to a perimeter of roughly ten miles distance from downtown. Areas beyond ten miles eventually were settled, although at a slower rate. After 1940, new growth took place more outside the city than within it.

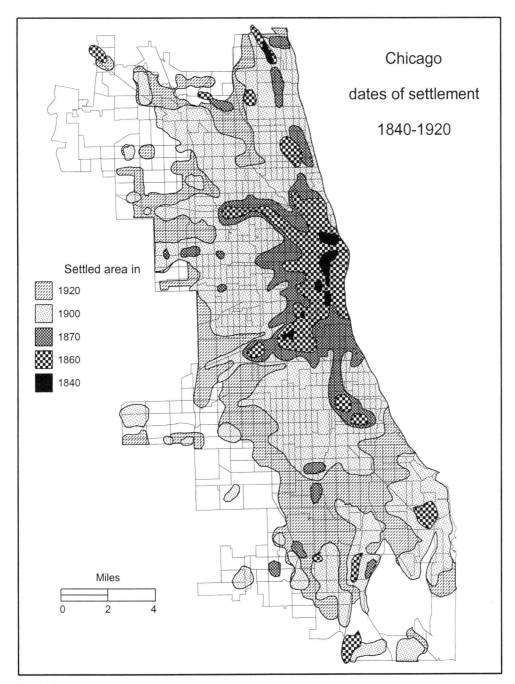

Fig. 11.1

Portrayal of growth as the outward spread of settlement from an original core offers one kind of perspective on the urbanization process. That is not the only dynamic that is involved, however, because older and more established areas of the

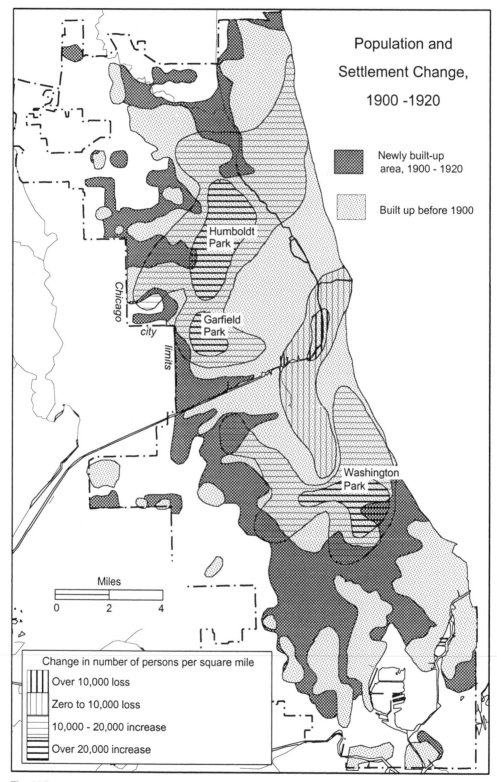

Population and Settlement Change, 1900 -1920

Newly built-up area, 1900 - 1920

Built up before 1900

Humboldt Park

Chicago

city

limits

Garfield Park

Washington Park

Miles

0   2   4

Change in number of persons per square mile

Over 10,000 loss

Zero to 10,000 loss

10,000 - 20,000 increase

Over 20,000 increase

Fig. 11.2

city also increased or decreased in population as the city grew over time. Data published by the Chicago Department of Development and Planning during the 1970s enable reconstruction of population changes on a square-mile basis for the period between 1900 and 1970.

Published census data are not as useful for longitudinal analysis because the areal reporting units changed several times during this period. In some years, populations are reported by townships or other minor civil divisions; in later years, wards of the city were used as the units. Local area data were published according to community areas in 1930, by wards again in 1940, and, beginning in 1950, only census tracts were used. Conversion of population data to a square-miles grid eliminates the problem of changing areal units.

The 1900-1920 period was one of intensive residential development in many parts of the city (Fig. 11.2). In terms of architectural styles, this was the "bungalow era," and most areas that experienced housing growth at that time still retain thousands of these attractive, one-and-a-half story houses, often built of brick and incorporating one or more porches in their design. The intensification of residential densities was most pronounced in a broad, semicircular band roughly four to six miles from downtown. The Humboldt Park, Garfield Park, and Washington Park neighborhoods were converted from low-density urban fringe zones to fairly high-density urban neighborhoods during the 1900-1920 period. These three parks and their connecting boulevards had been a central feature of the Olmsted and Vaux plans of 1870. The three neighborhoods became attractive sites for residential growth some three to four decades after their parks were set aside, no doubt much in the manner that Olmsted and Vaux had intended.

The overall pattern of population and settlement change between 1900 and 1920 shows a steadily advancing zone of new settlement centered approximately eight to ten miles out from the city center, an intermediate zone of intensifying residential densities from four to six miles out, and a zone of general population decrease in the older parts of the city. Commercial and industrial activities were crowding single-family residences out of most of the city center by that time.

ETHNIC CONCENTRATIONS

Within this overall dynamic were changes in population composition resulting from new patterns of both international and internal migration. In 1910, more than three-fourths of Chicago's residents either had been born in foreign countries or had at least one foreign-born parent. Germans accounted for the largest share of this foreign-born and foreign-stock population (Fig. 11.3). By 1910, Germans lived in all parts of the city, although their initial enclave between the North Branch of the Chicago River and Lake Michigan on the North Side continued to be the strongest

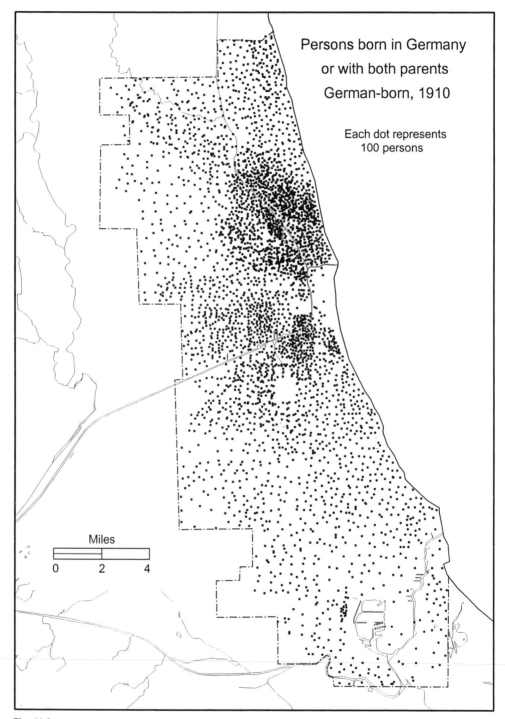

Persons born in Germany
or with both parents
German-born, 1910

Each dot represents
100 persons

Miles

0   2   4

Fig. 11.3

concentration. Second-generation Germans (native-born with German-born parents) moved out of those early neighborhoods and took up residence in parts of the city that had not existed when their parents had come to Chicago.

Another concentration of Germans is evident in what were then the city's ninth and tenth wards, west of the South Branch and south of Roosevelt Road. This neighborhood, which is known by various names, including Lower West Side and Pilsen, was home to more than 20,000 European immigrants in 1910. In this small area, ethnic Germans, German Jews, and others whose birthplace was classified as German were predominantly recent arrivals from Europe.

Germans were by far the city's largest foreign ethnic group. In 1910, they outnumbered the Irish-born and Irish-stock population in a ratio of two- or three-to-one over most of the city, even in the well-known Bridgeport neighborhood on the near Southwest Side (Fig. 11.4). The Irish came early, especially during the 1840s when famines at home sent thousands to the United States. Because Ireland's population was comparatively small, however, and because there was less need for emigration in the later years, the Irish were not replenished in substantial numbers in cities such as Chicago.

Like the Germans, the Irish-born lived in all parts of Chicago in 1910, although their areas of greatest concentration were in the older parts of the city. Irish populations also were established in the wards directly west and directly south of downtown. Both of those areas were swaths of territory that lay directly in the future expansion paths of African-American neighborhoods.

Chicago was home to some 45,000 Italian immigrants in 1910. Most of them lived between Van Buren Street and Roosevelt Road, east of Halsted (Fig. 11.5). This area, sometimes called the "Halsted slum," was a true immigrant ghetto in 1910. Russian and Polish Jews lived just to the south, between Roosevelt Road (12th Street) and 16th Street, a neighborhood that included the famed Maxwell Street market (Fig. 11.6). These were the ninth and tenth wards of the city, which housed nearly 100,000 people in a little over two square miles. In 1910, ninety-six percent of the area's residents were foreign-born or had foreign-born parents. Immigrants from Hungary, Greece, and Poland lived in the same general area.

Efforts to ameliorate conditions in the Halsted slum led to establishment of the Hull House, which was founded in 1889 by Jane Addams and Helen Gates Starr. Hull House evolved into a thirteen-building complex that housed schools, day care, and other services by 1910. Despite such help, the area lingered as a slum, often housing the most recent immigrants to the city, until much of it was razed in the 1960s for construction of the new University of Illinois at Chicago campus.

Ethnicity, nationality, and national status are not one and the same, and their changing definitions mask trends in the population over time. Poland did not officially exist when the census of 1910 was taken. Poles are included under Austria,

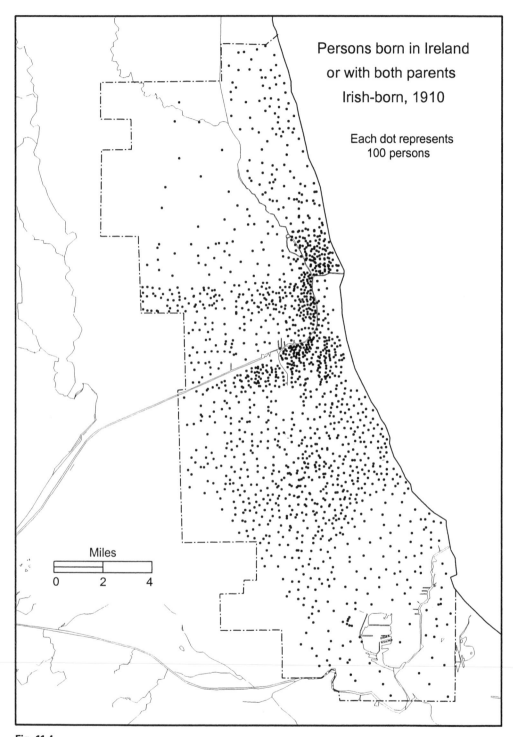

Persons born in Ireland
or with both parents
Irish-born, 1910

Each dot represents
100 persons

Miles
0   2   4

Fig. 11.4

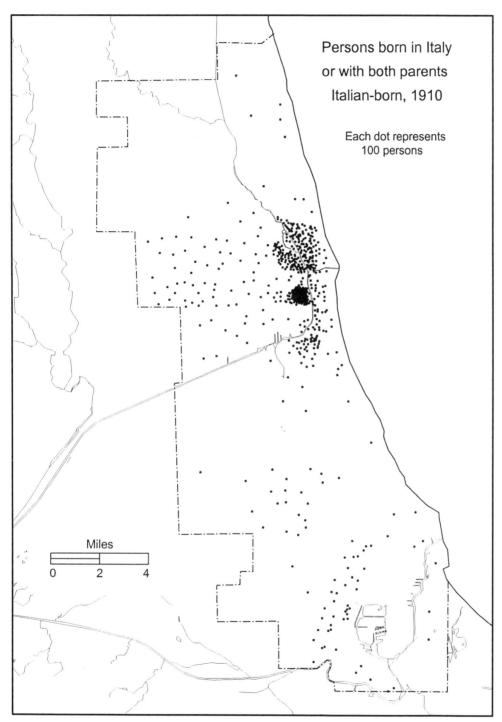

Fig. 11.5

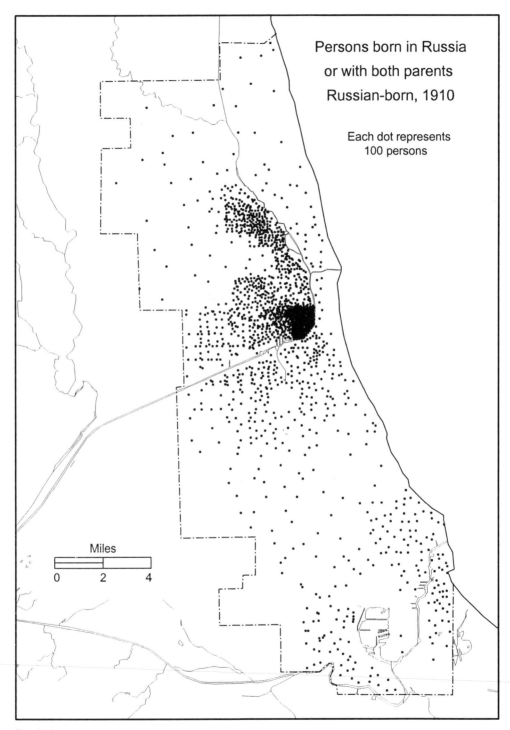

Persons born in Russia
or with both parents
Russian-born, 1910

Each dot represents
100 persons

Miles

0   2   4

Fig. 11.6

Germany, and Russia in the data for that year. Russian Jews, Polish Jews, and German Jews, while having affinities with one another, are disaggregated according to the countries in which they were born. While no map of the Polish population is possible for 1910, the pattern of Austrian-born residents in Chicago bears an unmistakable resemblance with the pattern of the Polish population in later years (Fig. 11.7).

Like the Germans before them, Polish people developed a strongly sectoral residential pattern. The first cluster, along Milwaukee Avenue east of Division Street, was evident even by 1860. Continued migration from Poland drew new arrivals to the old neighborhood, while the sons and daughters of the early Polish immigrants moved on toward the northwest, continuing the trend. For the next century and more, Polish populations would spread outward in an ever-expanding sectoral wedge.

Each ethnic group tended to produce its own geographical pattern within the city. Germans and Irish, who came first, had their neighborhood concentrations, but they soon spread throughout the city and rarely lived in inner-city slums. Italians and Russians (especially Russian Jews) came later and found shelter in neighborhoods that had been deteriorating for years. Their ethnic enclaves were at the high end of population density levels in the city, although in later years they, too, would disperse into various parts of the metropolitan area. Polish immigrants had an intermediate pattern of fairly discrete neighborhoods, but, because there was a continued migration from Poland in the following years, the first enclaves expanded outward from natural growth as well as immigration. Long-term fluctuations in the number of new migrants arriving in the city account for some, but not all, of these differences.

THE FIRST AFRICAN-AMERICAN NEIGHBORHOODS

None of these generalizations work very well to describe the evolution of African-American neighborhoods in Chicago. Chicago's black population grew from a few thousand in 1870 to more than 1.1 million in 1970. The first forty years of that century-long span saw only a modest growth in the number of African-Americans in the city. In 1900, their population stood at 30,000, roughly equal to the number of Canadians who lived in Chicago. But the tempo of change increased soon thereafter. Approximately 50,000 African-Americans moved to Chicago between 1910 and 1920. Escaping the declining fortunes of agriculture and the growing menace of white-on-black violence in the South, African-Americans began moving north in large numbers.

The problems were immediately obvious. Like all immigrant groups, African-Americans evolved their own institutions to cope with life in the city. The *Chicago Defender* newspaper, a mainstay of the black community, was founded in 1905. In 1919, race riots broke out in several northern cities, including Chicago. The first black neighborhood in Chicago was "Bronzeville," which grew on the southern edge of the downtown area. In 1910, roughly half of the city's 44,000 black inhabitants

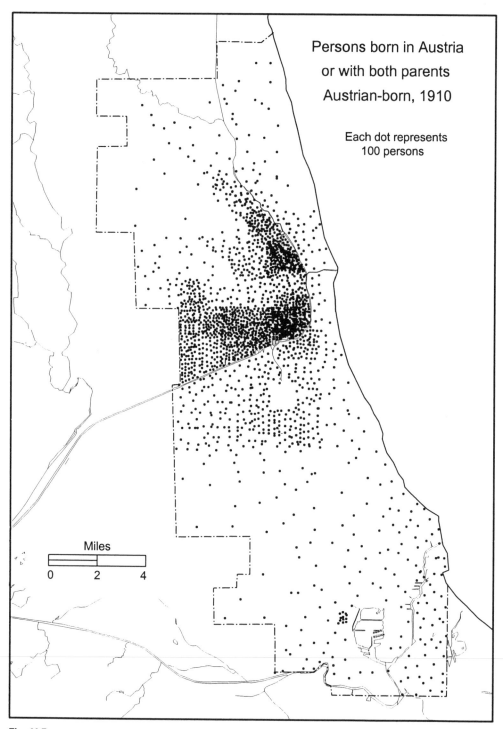

Persons born in Austria
or with both parents
Austrian-born, 1910

Each dot represents
100 persons

Miles

0   2   4

Fig. 11.7

lived between 22nd and 39th streets on the South Side, mostly east of State Street (Fig. 11.8). A much smaller concentration appeared on the West Side at the same time. It was out of these beginnings that the massive ghettoes of later years emerged.

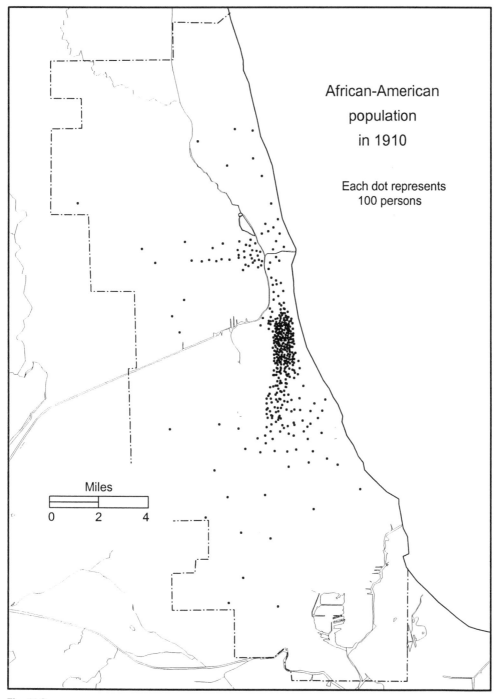

African-American
population
in 1910

Each dot represents
100 persons

Miles
0   2   4

Fig. 11.8

RESIDENTIAL GROWTH, 1920–1940

The overall pattern of population change in Chicago between 1920 and 1940 was a continuation of the trend established earlier in the century (Fig. 11.9). Downtown areas continued to experience a net population loss as residential land use expanded outward and was replaced by industrial or commercial activities near the city center. An intensive zone of residential expansion, four to six miles out from downtown, had characterized this period (Fig. 11.2). Between 1920 and 1940, the zone of most rapid development moved out to a semicircular zone of between eight and ten miles from the city center. Newly built-up areas were confined to the farthest extremities of the city after 1920.

A marked increase in population density accompanied conversion of single-family residential neighborhoods to apartment dwellings in the lakefront (Lake View) neighborhood on the North Side. Single-family residences accounted for most of the housing growth elsewhere, including the South Shore and Gage Park neighborhoods on the South Side, which experienced substantial population growth in the 1920s. Austin, Albany Park, and the Belmont-Cragin neighborhoods were counterparts of these growth areas on the North and West sides of the city. They were converted from low-density urban fringe settlements to single-family urban residential blocks. As on the South Side, most of the growth took place during the 1920s rather than the 1930s.

Population loss characterized a large area of central Chicago, even during a rapid growth decade such as the 1920s. In most cases the declines were due to land conversion, from residential to commercial or industrial uses. Such losses of inner-city population were common in the industrial cities of that era. Most housing that remained in industrial areas was of poor quality and was taken mainly by those who could afford no better. Twenty-year population losses of fifty percent or more characterized the industrial zones of the Near West Side at this time.

AFRICAN-AMERICAN POPULATION GROWTH TO 1950

The arrival of more African-American migrants sharply reversed the pattern of population loss in those parts of the central city where blacks were able to find a place to live. African-American populations doubled, tripled, and quadrupled in the wards south of downtown and east of State Street between 1910 and 1940. Older homes were converted to multifamily residences. Kitchenette-type apartments, designed to appeal to newly arrived black migrants, were built in the same areas. African-Americans accounted for seventy-four percent of the Douglas neighborhood's population by 1920 and made up eighty-eight percent of it ten years later. In Grand Boulevard, the population was ninety-five percent black in 1930 (Fig. 11.9).

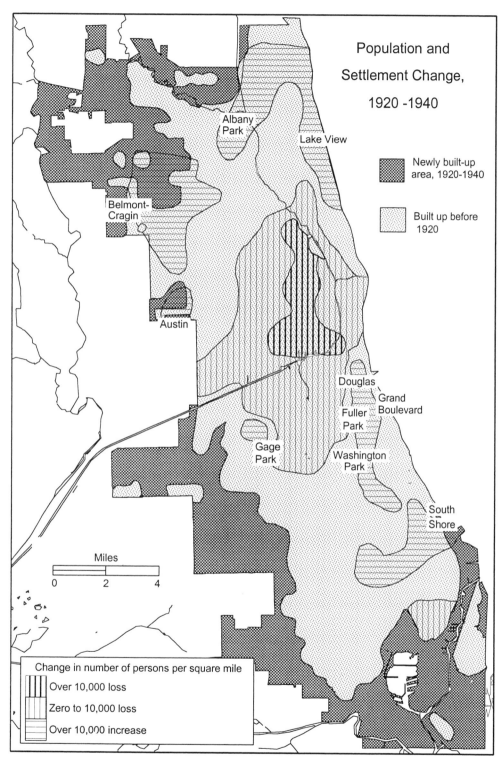

Population and Settlement Change, 1920 -1940

Albany Park

Lake View

Belmont-Cragin

Austin

Douglas

Grand Boulevard

Fuller Park

Gage Park

Washington Park

South Shore

Newly built-up area, 1920-1940

Built up before 1920

Miles

0   2   4

Change in number of persons per square mile

Over 10,000 loss

Zero to 10,000 loss

Over 10,000 increase

Fig. 11.9

Washington Park had been the scene of a residential subdivision and apartment construction boom around 1910. Separate residential areas contained about 32,000 whites and 6,000 African-Americans in 1920. The race riot of 1919 led many whites to move out of the neighborhood. Black families moved in to fill the vacancies, and by 1930 Washington Park was ninety-two percent African-American. Rapid change from white to black accompanied by a moderate to large increase in population density characterized neighborhood turnovers of this sort, which were typically accomplished within a decade.

By 1950, African-Americans were overwhelmingly the most important population subgroup from the edge of the downtown area south to 71st Street (Fig. 11.10). The concentration was heaviest in the mile-wide zone between State Street and Cottage Grove Avenue. Thirty census tracts in this elongated zone had more than 5,000 black residents each, and the total African-American population of the thirty tracts was 207,000. Fewer than 3,000 whites lived in the same area.

African-Americans also had moved into two areas directly west of downtown. Migrants from the Southern states first moved to the East Garfield Park neighborhood along Washington Boulevard in the 1930s. Although African-Americans made up only about fifteen percent of the local population in 1950, the small enclave grew rapidly in the following years as whites moved out. A second West Side area grew south of Roosevelt Road. Black migrants moved into areas that were being vacated by Eastern European Jews whose old neighborhood began to break up and disperse. These two early West Side African-American neighborhoods grew rapidly and coalesced by the 1960s.

A single outlier of black population was found on the North Side of Chicago in 1950. The Chicago Housing Authority built the Frances Cabrini Homes consisting of nearly 600 homes and apartments. Approximately 6,000 African-Americans lived in the area in 1950, forming the only significant concentration of black population north of downtown Chicago.

The incorporation of African-Americans into Chicago's population during the first half of the twentieth century was accompanied by changes unknown before that time. While several ethnic groups, including Italians and Eastern European Jews, had been strongly concentrated in limited areas, most of their immigration had taken place during a shorter time span and did not continue at high levels for more than one or two decades. African-Americans were severely limited as to where they might find housing, and the problem was compounded by the arrival of thousands more migrants every year. African-American population densities rivaled those of the most ghettoized European groups, but the total area of high-density black population was massively larger than anything known before.

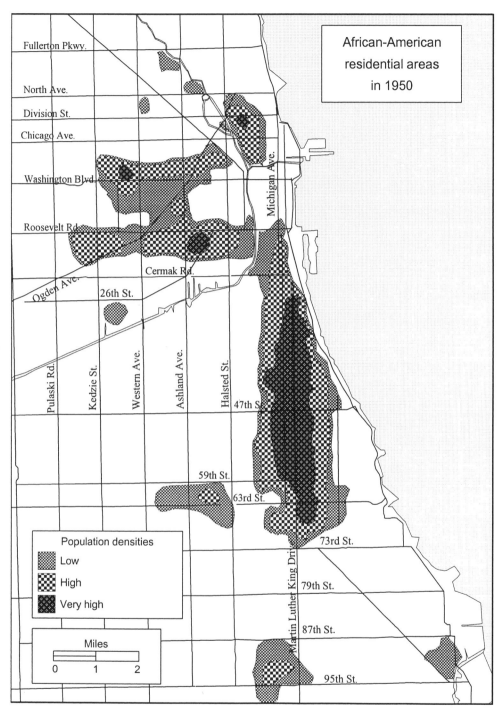

Fig. 11.10

The term "white flight," while it has emotional overtones and may be subject to misinterpretation, also has a degree of accuracy. The outmigration of European-derived peoples from areas adjacent to growing black neighborhoods resembled a flock of birds taking flight. The outmigration was rarely gradual and took place suddenly, typically leaving few behind. The process of white flight in advance of African-American residential expansion would become commonplace in Chicago after 1910, although it would not affect overall population growth in the city for several decades to come.

Chapter 12
TRANSPORTATION AND INDUSTRY

THE NETWORK OF EXPRESSWAYS AND INTERSTATE HIGHWAYS in and around Chicago is such a familiar part of the landscape it is hard to imagine that the entire system was created only during the past fifty years. Prior to the Interstate and Defense Highway Act of 1956, even large cities such as Chicago had very few miles of limited-access highways. The best available routes were the federal highways (Fig. 12.1).

Until the late 1950s, the closest approximation to a freeway bypass around Chicago was Mannheim Road, which carried traffic from a busy intersection in Des Plaines south through a built-up, urban-fringe zone. U.S. 30 (the Lincoln Highway) and U.S. 6 diverted cross-country traffic through downtown Joliet instead of through downtown Chicago. The bypass routes generally crossed other highways at four-way intersections, which produced lengthy delays to traffic. Within Chicago, both 12th Street (Roosevelt Road) and 95th Street were routes for cross-country travel, although both also carried heavy flows of local traffic. U.S. 41 and Alternate 30 followed winding routes through and around Jackson Park. Most of these routes still exist in this form today, but federal highways are no longer the only routes for long-distance travel.

Only a few of these roads resembled the radiating diagonals envisioned in the Burnham plan (Fig. 10.3). Most were upgraded segments of older, grid-line roads that were widened and straightened. Although the federal highways crossed state lines, the States of Indiana and Illinois had different ideas about the alignment routes should follow, one result being that the highways made jogs northward or southward near the state line to connect with their continuations on the other side.

INDUSTRIAL PATTERNS OF THE 1950s
The sparse grid of federal highways would have been totally inadequate had trucks been as important in interstate commerce then as they would become once the Inter-

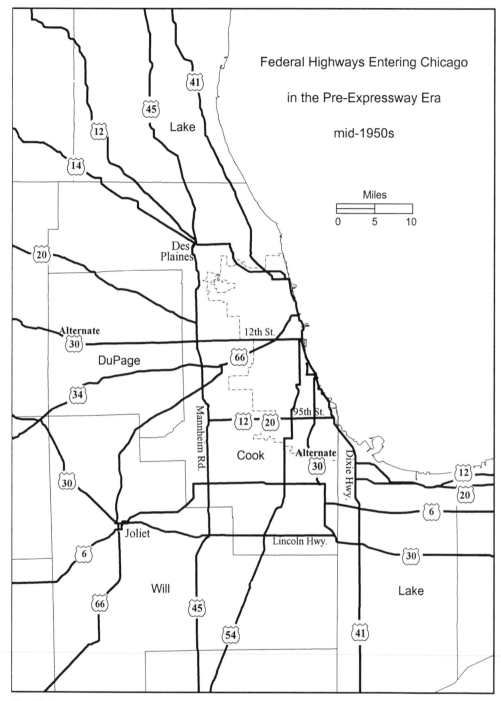

Fig. 12.1

state highway system was in place. Nearly all goods shipped to and from manufacturing plants still moved by rail in the 1950s. At the local scale, this meant that factory locations were bound to railroad corridors (Fig. 12.2). Although only a few highways

Fig. 12.2

encircled the city in beltway fashion, beginning in the 1900s railroad companies constructed their own versions of cutoffs and bypasses around Chicago, and these, in
turn, became zones in which new manufacturing plants were built in later years.

Most of the belt railway routes were short-distance segments designed more to provide access to local industries than to carry long-distance traffic. The industrial zones they spawned helped pull manufacturing jobs into the suburbs which, in turn, stimulated population growth in nearby areas. The pattern of manufacturing plants was not merely a reflection of the railway network. Railroad companies often were directly involved in developing new "industrial parks" along their lines, and they recruited firms seeking to build new facilities. Half a dozen or more manufacturing plants often shared the same railway spur line.

The circumferential railways also used the beltlines to connect with one another's yards to permit the interchange of freight traffic passing through (rather than originating or terminating) in the Chicago area. The largest of these railroad-corridor industrial zones was the Clearing district south of Midway Airport. A three-mile strip of manufacturing plants and warehouses, constituting the industrial suburb of Bedford Park, was built along the south side of 65th Street adjacent to the Chicago city limits. Clearing Yard, the largest of Chicago's freight interchange facilities, bordered the Bedford Park development on the south. More manufacturing plants and warehouses were added between the yard and 75th Street in later years. Bedford Park and Clearing were closely linked to another new industrial area at Summit (Argo) where a beltline railway crossed the Sanitary and Ship Canal. Corn milling and refining industries were concentrated there.

In the first half of the twentieth century, more industrial zones appeared on the South Side near Chicago Heights and Blue Island; at Franklin Park on the West Side; and in Skokie and Evanston on the North Side. All of these developments were guided by railroad real estate promoters. Industrial parks concentrated business conveniently for railroad companies. The practice also clustered plants and warehouses in a manner that reduced sprawl and lowered infrastructure costs by requiring fewer miles of streets and utilities.

Industrial expansion in the Calumet region of northwest Indiana focused on the heavy industries, especially steel manufacturing and petroleum refining (Fig. 12.3). The large complex of refineries at Whiting, Indiana, became the terminus for natural gas and oil pipelines built north from Texas and the Gulf Coast during the 1940s and 1950s. Steel manufacturing in the Calumet region reached a high point during World War II and the years immediately thereafter. As the industry modernized, it also downsized in response to competition from steel producers elsewhere. New steel mills continued to be built into the 1970s, however, including the complex of furnaces and rolling mills constructed at Burns Harbor. Completion of the St. Lawrence Seaway in the 1950s led to the construction of new lake port facilities, including the Port of Indiana. The port is a state-owned facility used to ship and receive grain, cement, and other bulk commodities.

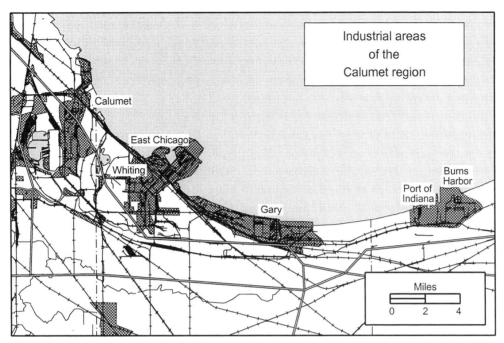

Fig. 12.3

By some measures, the late 1950s was the high point of Chicago's role as an industrial city. The downsizing, restructuring, job loss, and job migration that would characterize so many industries in later years had not yet begun. In 1958, the Chicago metropolitan area had more than 13,000 manufacturing plants, three-fourths of which were located in the city itself (Fig. 12.4). More than 500,000 production workers were employed in manufacturing. Nearly 50,000 were employed in the steel industries and three times that many workers took the products of steel mills and fabricated them into machinery and heavy-duty equipment. These were the industries that had long been concentrated in the Calumet region or on the South Side of the city.

The manufacture of electrical appliances, telephones, and radios required a labor force of 130,000 in 1958. The giant Western Electric manufacturing subsidiary of Bell Telephone employed thousands in its Hawthorne plant, which occupied several city blocks on the southeast corner of Cicero Avenue and 22nd Street (Cermak Road). Twenty thousand Chicago-area workers assembled radios and television sets in that era before production of these items moved overseas. The printing and publishing industries employed 80,000. The apparel industries had more than 30,000 production workers.

Food processing and preparation was another large Chicago industry in 1958, requiring more than 60,000 production workers. Meat packing and grain milling,

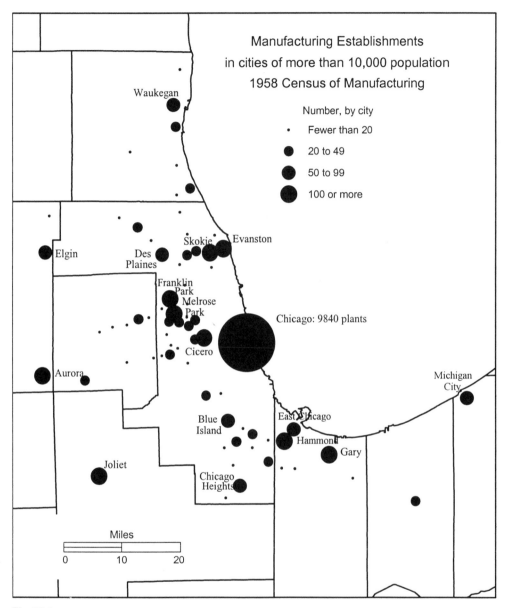

Manufacturing Establishments in cities of more than 10,000 population 1958 Census of Manufacturing

Number, by city

· Fewer than 20
● 20 to 49
● 50 to 99
● 100 or more

Waukegan

Elgin

Des Plaines

Skokie Evanston

Franklin Park

Melrose Park

Cicero

Chicago: 9840 plants

Aurora

Michigan City

Blue Island

East Chicago

Hammond Gary

Joliet

Chicago Heights

Miles

0   10   20

Fig. 12.4

two of the city's earliest industries, remained important. New uses for corn, especially the substitution of corn syrup for cane sugar as a sweetener, had made Chicago the center of the candy industry. Ready-to-eat foods of all types were processed and packaged in the city and suburbs. Cookies, crackers, candy bars, soft drinks, and breakfast foods were made in Chicago and shipped in all directions.

Although not a manufacturing activity as such, the mail-order business was concentrated in industrial areas of Chicago, and it used the same network of transportation routes to receive and ship goods. Sears, Roebuck and Company had its warehouses near Goose Island, north of downtown. Montgomery Ward and Company occupied a massive warehouse in the Central industrial district along 39th Street near the Union Stock Yards (Fig. 12.2). Built just after World War I, Central was one of the earliest planned industrial districts in any American city. Most of its factory and warehouse buildings remain in use today.

Even closer to the heart of the city, and built atop the railroad tracks, was the Merchandise Mart. It was constructed by Marshall Field and Co. on the north side of the Chicago River, at the junction of the river's two branches, in the mid-1920s. Its location in the very heart of Chicago—the most central place imaginable in the 1920s—reflected the pivotal role it was intended to play in the city's wholesale trade.

By the late 1950s, manufacturing industries had located in many of Chicago's suburbs. Industries found in the outlying clusters included metals fabricating, food processing, electrical machinery production, printing, and a variety of types of light manufacturing. Although the overall pattern of industrial sites was guided by the availability of railroad freight lines, which were practically ubiquitous, an increasing share of the plants were located in the western and northern suburbs. Most of them were new factories which specialized in products that had not been invented when Chicago's older industrial districts were established. Highway access also was important, especially to the smaller companies, which often preferred to ship their finished products by truck, even though they continued to receive raw materials by rail. Access to air freight transportation, then in its infancy, was important to some of the firms and they valued being close to O'Hare Field (O'Hare International Airport).

Value added by manufacture (selling price of the goods minus the cost of raw materials) is another measure of industrial activity (Fig. 12.5). It reflects both the profits made by companies and the wages paid to workers. The pattern of value added in 1958 is skewed toward the northern and western suburbs of the city. Although steel making and heavy machinery industries form a cluster of high value-added manufacturing in the South Side and Calumet region suburbs, the ring of satellite communities from Cicero to Evanston stands out even more clearly. Franklin Park, a largely industrial suburb, had nearly eighty small- to medium-sized factories in 1958. Along with its neighboring communities of North Lake and Melrose Park, this cluster of suburbs employed 20,000 manufacturing production workers.

The gradual drift of high value-added manufacturing toward the northwest portion of the Chicago metropolitan area was already underway by the 1950s. An even more pronounced shift in this direction awaited the coming of two changes in the

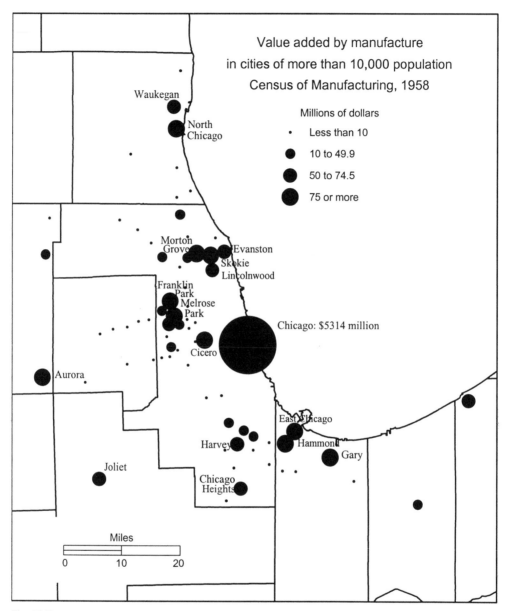

Fig. 12.5

transportation sector that were only then beginning to emerge. One was the development of Chicago's expressway system as part of the new Interstate highway network. The other was the advent of jet air transportation and the shift in service from Midway Airport to O'Hare. Over the next thirty years, the O'Hare Airport vicinity would eclipse downtown Chicago as the most accessible place in the metropolitan area.

O'Hare originated as a manufacturing plant rather than as an airport. With a large wartime defense contract, Douglas Aircraft Corporation began construction in 1942 of an aircraft factory and landing field on 1,000 acres of former farm and orchard land a few miles northwest of the Chicago city limits. The plant, which employed more than 20,000 workers at the peak of wartime production, eventually produced hundreds of military aircraft before it was closed at the end of World War II. The federal government then gave the facility to the City of Chicago. The city annexed a strip of land around it and acquired a narrow corridor connecting the airport to the city. O'Hare Field (named for World War II flying hero, Edward O'Hare) became a commercial airport in 1949.

Midway Airport was the favored hub of commercial air travel during the piston aircraft era, and it remained in that role even after jet air travel became important in the 1960s. Like O'Hare, Midway Airport had been built on the edge of the city; but Midway was bordered by residential and industrial developments that prevented construction of the longer runways needed for the larger jet aircraft. O'Hare had the necessary room for expansion, and it soon became the hub of traffic for Chicago. Midway's eclipse eventually led to its closure as a terminal for regularly scheduled passenger travel, although by the mid-1980s it was reopened as the demand for air travel continued to grow.

THE EXPRESSWAY SYSTEM

Development of the O'Hare facility preceded construction of Chicago's Interstate highways, but only by a few years. The first sections of Edens Expressway were built on the northwest side of Chicago during the mid-1950s (Fig. 12.6). Within five years, the major expressway routes through and around the city either had been completed or were under construction. Interstate highways 90 and 94 were the first to be completed and were soon complemented by Interstate 294, the circumferential freeway bypass. Interstate 80, which followed the east-west alignments of U.S. highways 6 and 30, was added a few years later. Segments were added more slowly after that time, but the few initial spokes have been augmented in the years since until the present-day web-like network of expressway routes was in place.

Most of the new Interstates followed the alignment of existing federal highways, although there were a few significant new routes. The Calumet Skyway, a ten-mile-long series of bridges and overpasses, carried traffic over and above the busy Calumet region's steel mills and refineries. The Dan Ryan Expressway was built on a north-south alignment for ninety-five blocks, paralleling State Street. Construction of the Dan Ryan and the Congress Street (later named Eisenhower) Expressway directly west from downtown Chicago produced massive disruption of residential areas, many of which were considered ripe for urban renewal.

Fig. 12.6

Originally known as "super highways," the new system of limited access express-ways was soon found to be inadequate. The amount of new traffic the highways stimulated was not anticipated when they were being planned and built. New lanes were added, some existing lanes were signaled for use in either direction of traffic flow to accommodate rush-hour traffic, and both the Northwest and Dan Ryan expressways had rapid-transit rail routes built down their median strips.

The effect of the superhighway was to obliterate any existing land uses on or adjoining its right-of-way. The early expressways had routinely destroyed "blighted" inner-city neighborhoods in the spirit of slum clearance and urban renewal, but by the 1960s local residents often united in protest when any new expressway route was proposed. The escalating costs of freeway construction, the protests inevitably resulting from the selection of any particular route, and the knowledge that new highways would merely stimulate more traffic brought an end to the era of urban freeway building by the 1970s.

By that time, the growth of O'Hare Airport and the reconfiguration of traffic flows within and around Chicago had caused an apparently permanent shift in the city's economic structure. The Loop was no longer the most accessible part of the city; the I-90/I-294 highway junction near O'Hare had replaced it. Although hundreds of thousands of jobs—particularly white-collar jobs—remained in downtown Chicago, the creation of new jobs was more closely linked to sites that had freeway access and were conveniently close to the airport. Retail-store chains built new facilities in the outlying shopping centers. Added to these factors was the gradual migration of population toward the northern and western suburbs of the city, a growth that eventually led to the creation of more jobs on the city's northwest fringe.

INDUSTRIAL PATTERNS IN THE 1990s

By 1997, manufacturing industries had dispersed widely over the suburbs (Fig. 12.7). (Some firms are not shown because they are located in cities of fewer than 10,000 population and are not represented in the published census statistics.) The northwestward drift of manufacturing first seen in the 1950s is even more evident on the 1997 map. The largest manufacturing suburb is Elk Grove Village, adjacent to O'Hare Airport, which employs more than 30,000 workers. Libertyville, Addison, Wheeling, and Arlington Heights also had become major manufacturing centers by the 1990s, even though their populations are comparatively small. In all, some 200,000 workers are employed in manufacturing in the firms located within a dozen miles from O'Hare. Chicago's manufacturing sector has declined in importance and so has that of many of the older suburbs. The Calumet region has dwindled significantly in this respect.

Both the building of O'Hare and the construction of an expressway system had the effect of unbalancing what had long been a remarkably symmetrical residential and land-use pattern in Chicago. The symmetries of times past depended on a single downtown focus that would be the universal point of access for all who lived in or in some way used the city. As this single, central focus weakened, it was replaced by a more uncertain center, at one or perhaps several locations on the periphery. Added to

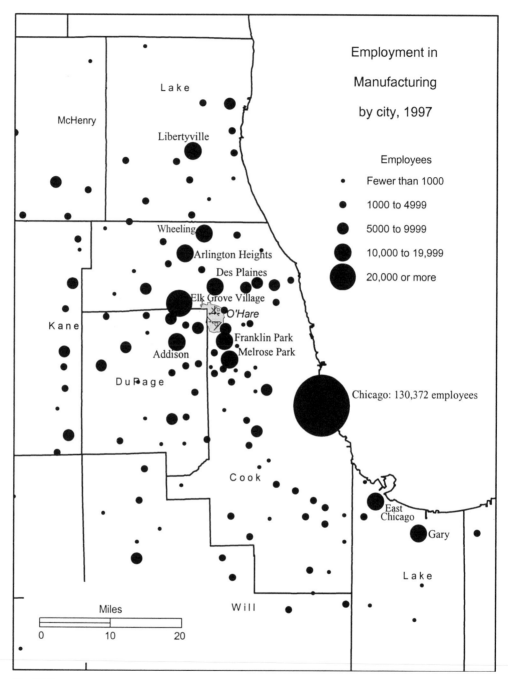

Fig. 12.7

these effects was a powerful dynamic of population turnover and neighborhood change within the city that eventually led to a net relocation of population to the northwest. These were developments of the "migration decades," the period from the close of World War II until roughly the mid-1970s.

Part V

THE CHANGING CITY

Chapter 13
THE MIGRATION DECADES

If the World's Fair of 1893 and the years of inspired city planning both leading up to and following it are considered the high point of Chicago's history, then the decades following World War II could equally be regarded as the low. These were the migration decades, which saw a massive exodus of the city's white population and its replacement by African-Americans. While migration itself has no negative connotation, the forces that drove migration during the postwar decades had a destructive effect on the city. Racism on the part of whites was revealed in their willingness to abandon the neighborhoods in which they lived rather than accept the presence of African-Americans. Rural black migrants from the Southern states lacked job skills and were unprepared for the type of life they would find in Chicago.

THE PROCESS OF NEIGHBORHOOD TURNOVER

Although there seems to have been no precedent for it—whether in Chicago or any other American city—a cycle of neighborhood change, driven by fear and racial prejudice, evolved rapidly as more African-Americans came to the North. "White flight" emerged by 1920, but it became more prevalent following World War II and was an expected part of residential patterns soon thereafter. In Chicago, the same scenario of population turnover was repeated in neighborhood after neighborhood, especially on the South and West sides of the city. Real estate agents attempted to steer African-Americans toward established black enclaves rather than aid their entry into all-white areas. Mortgage-lending institutions typically cooperated by unnecessarily delaying or refusing to make home loans that would change the existing color pattern. White home owners actively and visibly proclaimed their unwillingness to accept any African-American families as neighbors.

The process continued until at least a few whites, fearing the collapse of property values that surely would follow, sold their homes and moved elsewhere while real estate values still remained high. Once the first African-Americans moved in and property values sank rapidly, the remaining whites sold their homes at whatever price they could obtain, and the neighborhood became almost entirely black. This was the "tipping point," a term coined by social scientists to determine the amount of black migration necessary to start the wave of panic selling and mass exodus that turned an all-white neighborhood into an all-black one. While no such number ever could be specified, it was not really necessary to know. People had learned that the result was inevitable: as long as African-American population growth continued in the city as a whole, black neighborhoods would continue to expand, most often into adjacent areas.

The result was the replacement of one population by another, usually accompanied by a reduction in the total population. The early years of African-American migration to Chicago had produced an increase in population densities in the wards immediately south of downtown, where blacks crowded into older single-family homes that had been converted to multifamily residences. As the tempo of migration increased in later decades, little attention was given to remodeling the structures abandoned by whites as they fled. A period of disinvestment typically preceded the arrival of black migrants, which meant that African-Americans inherited the same housing stock, only now in poorer condition than before. Some houses and apartment buildings were abandoned outright while others were in such poor condition that they were soon deserted.

The magnitude and nature of African-American population growth changed almost constantly during the migration decades. Chicago's black population grew from 277,231 in 1940 to the peak census year, 1980, when 1,187,905 African-Americans were recorded in the city. Growth was not steady during these four decades, however, and the source of the increase changed. Natural increase replaced migration as the major source of black population growth by the 1960s. Eventually black outmigration increased to the level that it overwhelmed natural increase, which, of course, produced an overall net decline in the African-American population.

The major impetus to growth in the 1940s was the large influx of African-Americans that took place following World War II. Many of the new arrivals were young and came from rural backgrounds in the Southern states. More than half the increase in black population during the 1940s was caused by a net immigration of persons in the 19 to 35 age group. Because of their ages, most of the migrants came in search of work. Women slightly outnumbered men in this migration, especially in the 15 to 24 age range.

The best housing open to them was on the fringes of the existing black neighborhoods, on both the South and West sides of the city. White flight was evident immediately (Fig. 13.1). The map showing white outmigration by decade is based on a

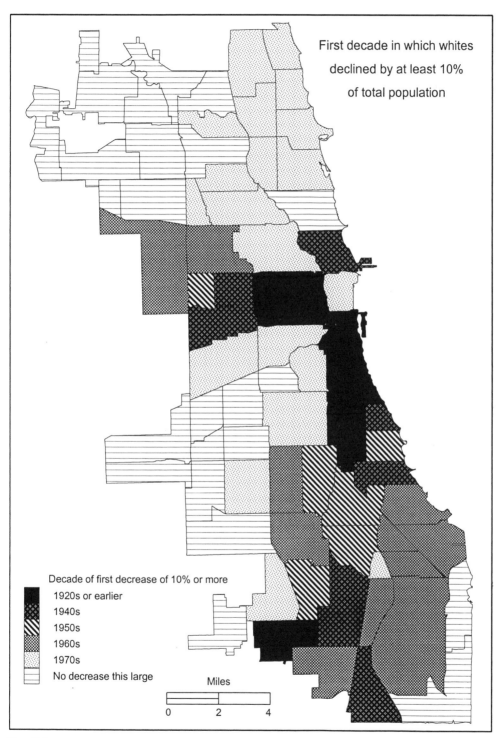

First decade in which whites
declined by at least 10%
of total population

Decade of first decrease of 10% or more

1920s or earlier

1940s

1950s

1960s

1970s

No decrease this large

Miles

0   2   4

Fig. 13.1

decline of ten percent in the white population. While a ten-percent reduction does not qualify as "flight," no other factors triggered white population losses of even this magnitude in what had been a growing city. Many neighborhoods lost white population even more rapidly once the ten-percent level was reached. The proportion of whites eventually declined to nearly zero.

ETHNIC AND RACIAL COMPOSITION OF NEIGHBORHOODS

Like all processes of neighborhood change, the arrival of African-Americans had been preceded by the arrival of other groups. In 1950, thousands of Italians were living in the central and western portions of the city (Fig. 13.2). Many had moved there only a few decades earlier, fleeing the densely settled Italian enclave that had grown up just west of downtown in the early decades of the twentieth century (Fig. 11.5). Italians moved west to the Humboldt Park and East Garfield Park neighborhoods during the 1930s. Italians also had a substantial presence in neighborhoods directly north and directly south of downtown, both of which were areas that African-Americans would soon enter.

Russians had been crowded into a small area of poor housing, just north of the inner-city Italian neighborhood, in the early twentieth century. They moved west, to the Garfield Park and North Lawndale neighborhoods, beginning in the 1930s (Fig. 13.3). Along with the Italians, the Russians—many of Jewish ancestry—constituted the majority foreign-born presence in those areas. By 1950, many West Side Jewish residents had begun to relocate again, this time northward to the Albany Park and Rogers Park neighborhoods. Thousands of Jewish refugees from Europe, including many Holocaust survivors, swelled the ranks of this group after World War II. They joined the migration to the North Side neighborhoods and, soon thereafter, to the new suburb of Skokie just north of the Chicago city limit. North Lawndale's large Jewish community was short-lived, but its population turnover was one of the most rapid anywhere in the city. In 1950, North Lawndale's population was eighty-seven percent white; by 1960, it was ninety-one percent black (Fig. 13.4).

Other rapid turnovers included the blue-collar community of Riverdale on the Far South Side. This area, along the Little Calumet River, was adjacent to a zone of steel mills and other heavy industries. Riverdale's population was 99.6 percent white in 1940. During World War II, the Chicago Housing Authority constructed the Altgeld Gardens housing complex in Riverdale, and many of its first occupants were African-Americans. The public housing project was expanded in the 1950s, and, again, many of those who found residence there were black. Although the public housing developments did not cover all of Riverdale, the resident white population moved out rapidly in the late 1940s. By 1950, Riverdale was 84.1 percent black.

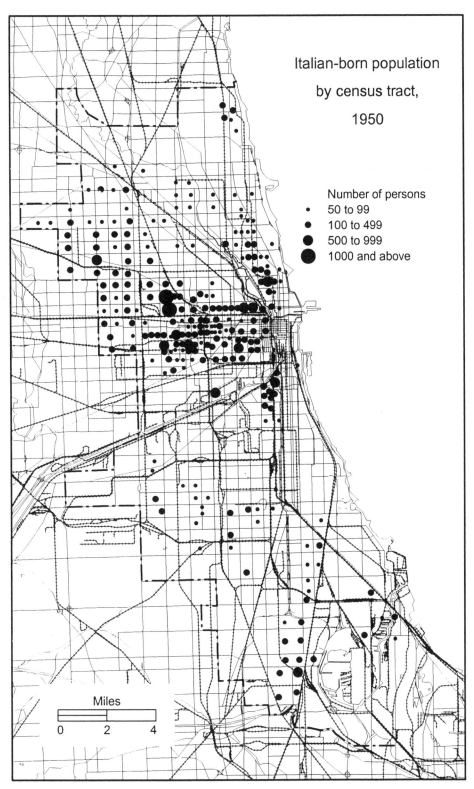

Italian-born population by census tract, 1950

Number of persons
· 50 to 99
• 100 to 499
● 500 to 999
⬤ 1000 and above

Miles
0   2   4

Fig. 13.2

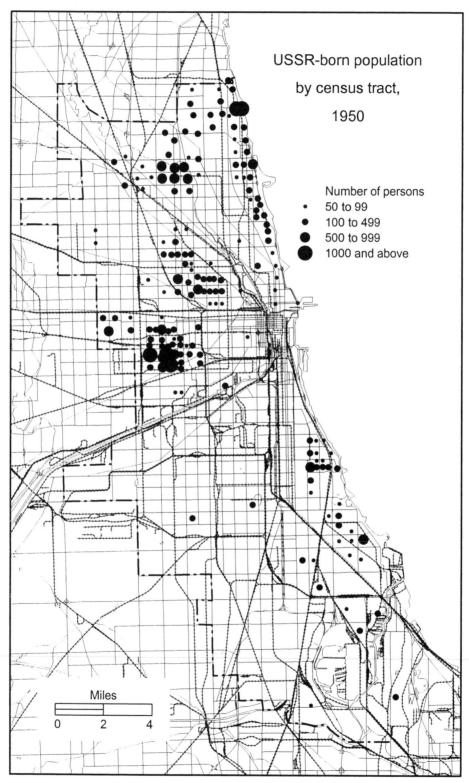

USSR-born population
by census tract,
1950

Number of persons
· 50 to 99
● 100 to 499
● 500 to 999
● 1000 and above

Miles
0 2 4

Fig. 13.3

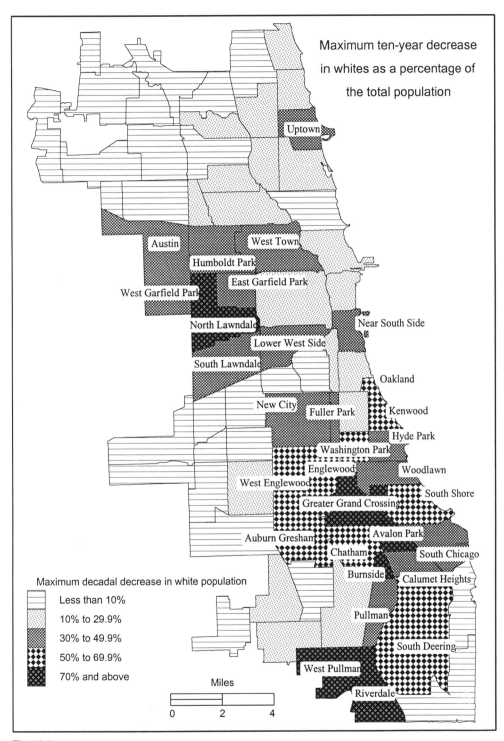

Maximum ten-year decrease in whites as a percentage of the total population

Uptown

Austin

West Town

Humboldt Park

East Garfield Park

West Garfield Park

North Lawndale

Near South Side

Lower West Side

South Lawndale

Oakland

New City

Fuller Park

Kenwood

Hyde Park

Washington Park

Englewood

Woodlawn

West Englewood

South Shore

Greater Grand Crossing

Auburn Gresham

Avalon Park

Chatham

South Chicago

Burnside

Calumet Heights

Pullman

South Deering

West Pullman

Riverdale

Maximum decadal decrease in white population

Less than 10%

10% to 29.9%

30% to 49.9%

50% to 69.9%

70% and above

Miles

0   2   4

Fig. 13.4

A similar sequence took place in the Greater Grand Crossing neighborhood. No single ethnic group constituted an overwhelming presence before 1950, although substantial numbers of Italian-, Irish-, German-, and Swedish-Americans lived there. African-Americans were few before the 1940s; but, as the postwar growth phase began, blacks moved into Greater Grand Crossing's apartments and single-family homes in substantial numbers. The community was 94.1 percent white in 1950; by 1960, Grand Crossing was 85.8 percent black.

The actual numbers of people represented in these mass movements were considerable because they involved an almost total turnover of population. Grand Crossing's upheaval saw the outmigration of roughly 50,000 whites and the in-migration of 50,000 blacks, all in one decade. In North Lawndale, the turnover amounted to twice that many people leaving and entering in a single ten-year period. Neighborhood populations typically dropped after the change took place, and overall depopulation eventually became the norm.

Other community areas attempted to control the changes that, left unchecked, would lead to total neighborhood turnover. During the 1950s, African-Americans moved into the Hyde Park neighborhood in significant numbers. White population dropped from 94.9 percent of the total in 1950 to 59.7 percent in 1960. A coalition of area residents formed the Hyde Park-Kenwood Community Conference to integrate the community racially without causing further white population loss. Funds provided by the University of Chicago and several major corporations supported various community redevelopment efforts, including the construction of new private and public housing. At the neighborhood scale, Hyde Park's population has achieved a stable balance between white and black. Individual census tracts within the neighborhood are more homogeneous, either white or black.

EXPANSION OF AFRICAN-AMERICAN NEIGHBORHOODS AFTER 1970

The 1960s was the decade of greatest growth in Chicago's African-American population, although in-migration had fallen off considerably to the point that it accounted for only about one-fifth of the growth. The rest was due to natural increase. The fertility rate (number of children under age five per woman ages 15 to 44) can be used to measure the reproductive rate of the two populations. Although black fertility rates often are higher than for whites, this was not true in Chicago until the 1950s. Mainly because so many of the pre-1950 migrants were young and had not yet started families, black fertility rates were below those of whites until the "baby boom" of the 1950s. From that time onward, child-bearing rates fluctuated up and down, decade by decade. In 1970, the fertility rate was .362 children under five per woman ages 15 to 44 among African-Americans; the same statistic was .304 children for white women in the child-bearing years. Black fertility was higher, but the difference was not great.

The 1960s was the last decade in which substantial numbers of African-Americans moved to Chicago from other places. The city's black population was still growing from net migration, child-bearing had increased for blacks (as it had for whites), and black outmigration was not yet a factor. African-Americans moved into neighborhoods on the Far Southeast Side of Chicago and into several on the Far West Side, including the Austin neighborhood where blacks met bitter opposition. The Austin Community Organization and several other groups were organized in an attempt to keep the neighborhood's racial composition from changing. The efforts soon failed, however, because the organized white population itself began to dwindle by the end of the 1960s. Not one African-American was recorded as living in Austin in 1960, but the population was two-thirds black by 1970.

The specter of white violence and intimidation of black families mingled in the 1960s with equally unwelcome images of black violence and destruction in their own neighborhoods. Chicago had a long history of violent uprisings in its black neighborhoods, and the 1960s was no exception. Even as these events transpired, however, the nature of African-American population dynamics was changing. The slowing rate of black inmigration was reversed to a net outflow in the 1970s. During that decade, the city showed a net loss of perhaps 75,000 African-Americans due to outmigration, even though the population still continued to grow from natural increase.

Some of the black outmigration was from the central city to the suburbs, following the same trend that was strongly evident in the white population. Some also was part of the "return migration" of that era, a nationwide trend in which significant numbers of African-Americans moved from Northern to Southern states. White flight from the central city continued, however, as did the incremental growth of the total area of the city occupied largely by African-Americans.

New areas occupied by African-Americans during the 1970s included a narrow fringe of census tracts on all sides of what had by then become a massive South Side black concentration (Fig. 13.5). Advances into the remainder of the city's West Side (and into adjacent suburbs such as Oak Park) took place at this time. African-Americans began moving to the city's North Side in significant numbers for the first time as well.

While it was often assumed that black migration and natural increase were continuing to produce the growth of black-only neighborhoods, white flight was obviously of equal importance in producing such a rapid turnover. Roughly two-fifths of Chicago's whites ages 20 to 29 in 1960 were not in the city when the 1970 census was taken. This is a massive population loss due to net migration that exceeds, for exam-

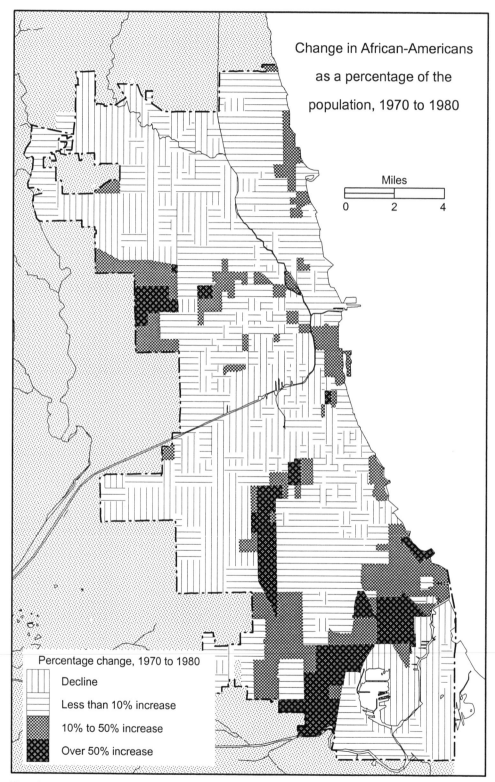

Change in African-Americans as a percentage of the population, 1970 to 1980

Miles
0   2   4

Percentage change, 1970 to 1980

Decline

Less than 10% increase

10% to 50% increase

Over 50% increase

Fig. 13.5

ple, the rate of leaving during conditions of the Dust Bowl in the Great Plains during the 1930s. White population losses through the 1960s were generally black population gains, although Hispanic migration to Chicago had become a factor by that time as well. Percentage gains in the local African-American population could equally well be produced by a black increase or a white population loss.

The hollowed-out shape of the expanding South Side black residential area during the 1970s is a key to understanding the process (Fig. 13.5). Even though Chicago was losing more African-Americans through migration than it was gaining, natural increase was sufficient to produce a small black population gain over much of the city. African-Americans living in the desperate housing conditions of the inner city were experiencing pressure to move. Public housing projects, such as the twenty-eight-building complex known as the Robert Taylor Homes on the Near South Side, were built to replace hopelessly dilapidated structures in traditional black neighborhoods. Not everyone qualified for public housing, however, and the number of new housing units in such developments rarely equaled the number lost through urban renewal.

The result was a constant pressure to move outward. White opposition channeled black expansion into fairly narrow zones adjacent to black neighborhoods. The combined effect of drastic losses in white population, a seemingly ever-expanding black zone on the South Side, plus overall black population increase, suggested that more of the same was to come. In fact, the entire process was about to collapse. Like their white predecessors had done a few decades before, African-Americans deserted Chicago by the tens of thousands during the 1980s.

The massive black ghetto of the inner city still exists, but it has been declining in population for three decades, and the downward trend is likely to continue (Fig. 13.6). Public housing that replaced dilapidated older housing in the 1950s and 1960s has itself been replaced. The high-rise apartment towers, a hallmark of public housing projects in the 1960s, have gradually been removed and replaced by smaller-scale clusters of public housing, built on a more human scale. On a time scale of thirty years, the only areas of the city that experienced a substantial net increase in African-American populations are some of those most recently occupied. Small percentage increases in black populations have taken place on the city's Northwest Side, but they appear as increases only because practically no African-Americans lived in such areas thirty years ago.

The 2000 census map of African-American population in Chicago and its immediately adjacent Illinois suburbs reveals some population movement into nearby communities, including Maywood and Melrose Park directly to the west (Fig. 13.7). Evanston's black community, which is of long standing and did not originate with

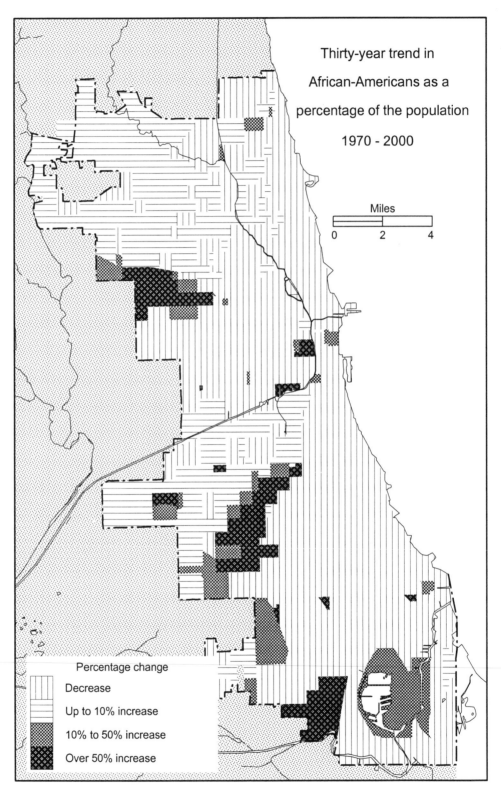

Thirty-year trend in
African-Americans as a
percentage of the population
1970 - 2000

Miles

0   2   4

Percentage change

Decrease

Up to 10% increase

10% to 50% increase

Over 50% increase

Fig. 13.6

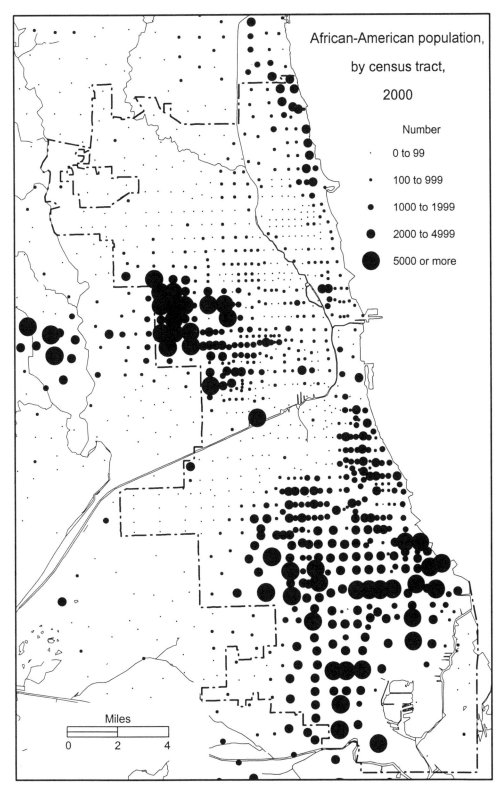

African-American population,
by census tract,
2000

Number

·   0 to 99

•   100 to 999

●   1000 to 1999

⬤   2000 to 4999

⬤   5000 or more

Miles

0   2   4

Fig. 13.7

suburban expansion, remains fairly small. Within Chicago, approximately 200,000 African-Americans lived in the West Side black neighborhoods in 2000, which is the highest concentration in terms of population density. Another 385,000 live between 63rd and 95th streets on the South Side. Only 125,000 African-Americans lived in the oldest of Chicago's black neighborhoods, between the South Loop and 55th Street, east of State Street.

The decay of inner-city housing, the experience of white flight, and a modest rate of growth in Chicago's black population through natural increase combined to drive African-Americans into new areas of the city for three decades after their migration from outside areas had dwindled to insignificance. A population expanding in area more rapidly than it expands in size often is in decline, numerically, where it once was growing. This is certainly true of Chicago's African-American population, which is now also declining in terms of absolute numbers. A small decrease in the number of African-Americans occurred in the 1990s, but the loss was offset by a small growth in the city's white population and a substantial growth in its Hispanic and Asian populations. With these changes, the migration decades came to an end as Chicago began to experience new trends in its pattern of residential neighborhoods.

Chapter 14
SUBURBAN EXPANSION

Nearly all American cities have developed a fringe of outlying suburbs during the course of their history. Chicago is no exception. It has spawned as many satellite communities as any large city, even Los Angeles. Chicago's early suburbs, like those of New York City, Boston, and Philadelphia, tended to be enclaves of the upper class. Those who could afford to live at some distance from the central city took advantage of their wealth and status by establishing homes—sometimes second homes—in the hilly, wooded, or lake-dotted countryside a comfortable distance away from the fast pace of urban life.

The circumstances that led to the creation of satellite communities varied over time. In the Chicago region, many suburbs began as small, independent trade-center towns in the countryside. They became suburbs when a railroad was built either through the community or close to it, which often reduced travel time to the city from hours to minutes. To take two examples in the Chicago area, Libertyville in Lake County, Illinois, was a small crossroads settlement until it was connected to the city by a railroad line in 1882. Libertyville's population grew rapidly for a time thereafter, and it became one of the Chicago's more attractive outer suburbs. Palatine had a similar history. Originally only a small crossroads trade center, Palatine became a village when a townsite was platted along the newly built Chicago and North Western Railroad in 1855. The village soon became home to a number of Chicagoans who appreciated its direct train service to the city.

Evanston and Lake Forest were selected as sites for two Methodist educational institutions in the 1850s. Both Northwestern University and Lake Forest College became the foci of new town developments adjacent to their sites. Naperville began as a trade-center town for a local farming community in the 1830s. It received a railroad in 1863 and in 1870 became the new home of North Central College, which was a factor in the community's growth thereafter. Naperville entered a different phase of growth in the 1980s when it acquired many more housing developments, shopping centers, and office parks. By 2000, Naperville, with a population of 128,000, was Chicago's largest suburb.

A few towns began with inspired designs for their future growth. Riverside, a western suburb of Chicago, was designed by Frederick Law Olmsted in 1869. Riverside's curving streets followed the contours of the land, similar to the design Olmsted and Calvert Vaux had created for New York City's Central Park. Lake Forest, which also was based on Olmsted's designs, had a similar layout. Both communities became enclaves for the upper classes.

At the other extreme were places such as Lansing, on the Little Calumet River. Lansing was the trade center for a population of Dutch and German farmers who raised produce on the rich muck lands of the glacial plain. Lansing acquired a railroad in 1856, but it did not become a commuter suburb. Population growth was slow in the area until steel mills were built in the Calumet region after 1900. The community became home to a largely working-class population that was employed in the nearby mills.

No two suburbs have identical histories, of course, and the factors that caused one to grow large while another languished varied a great deal from place to place. Until the automobile became a factor in human mobility in the 1920s, suburbs were connected to the central city by railroads alone. No other mode of transportation was significant in their growth. Even as late as 1950, the pattern of Chicago's suburbs essentially followed that of the radiating railroad lines (Fig. 14.1). In that year, an urban population of more than two million lived within a forty-mile radius of downtown Chicago; but more than three-fourths of those two million people also lived within ten miles of the city center. Pre-1950s suburbia was well supplied with communities of residence, and they were of various types.; but the number of people living in those communities was small compared with the concentration found in the central city.

SUBURBIA OF THE 1950s AND 1960s

This condition began to change during the 1950s, which was the first decade when significant construction of new housing developments took place at some distance from downtown. A pent-up demand for suburban housing had developed during the previous two decades. The 1920s had been a period of substantial new home construction, although much of it took place within the central city. Construction slowed markedly during the Great Depression, which was followed by World War II, another period of relatively slow housing growth. Difficulty in obtaining building

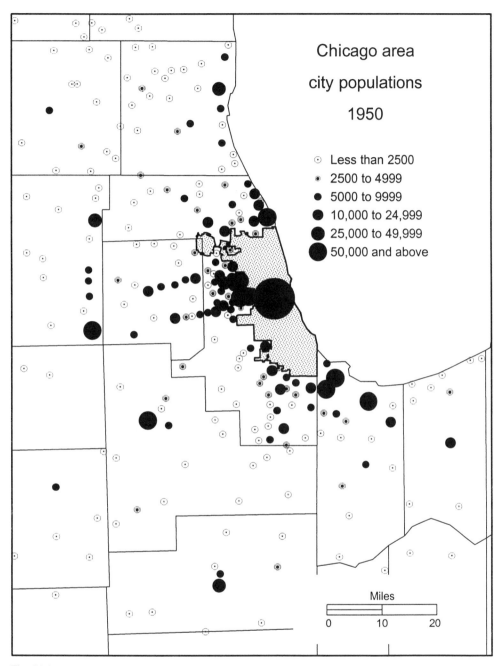

Fig. 14.1

materials and the general disruption of economic life that characterized the war years diverted attention away from new home construction. But when young men and women returned to civilian life beginning in 1945, there was an immediate demand for new housing. Suburban areas were to absorb most of the housing growth from that time forward (Fig. 14.2).

The maps portraying Chicago's suburban expansion are based on age of owner-occupied dwelling data for census tracts published in the 2000 Census of Population. Tracts shown as "built up during 1940s," for example, are those with a median age of dwelling ranging from 1940 to 1949. Most of the areas that were built up during the 1940s were either just within or immediately adjacent to the Chicago city limits. The far northwest corner of the city, including the Norwood Park, Edgebrook, and Forest Glen neighborhoods, received many new homes at that time.

Perhaps the most significant addition to Chicago's suburbs at the end of the 1940s was the new community of Park Forest, which was developed on the southern fringes of the metropolitan area as a planned community. Developer Philip Klutznick's organization, American Community Builders, Inc., built Park Forest along lines similar to other new communities of the period, such as the Levittowns constructed near Philadelphia and New York City. These communities were hallmarks of the new "suburbia" that characterized the postwar period. The old grid pattern of streets was cast aside in favor of curving streets, cul-de-sacs, and courts. Planned shopping centers were integrated into the community's design. Parks, schools, and other typical urban land uses were part of the package.

Klutznick's organization created housing for a variety of income levels, although social class was deemphasized (Whyte, 1956). Most of Park Forest's new inhabitants were young, white families. White-collar professional and technical occupations predominated. It was a new way of life, designed to appeal to young couples who were just starting families and desired to make a fresh start in a new community. These values would predominate in the design of many more housing developments around Chicago in the years to come, although few of them were as thoroughly planned as Park Forest.

The suburbs expanded in all parts of the Chicago metropolitan area during the 1950s. That decade also witnessed substantial new growth in the form of housing developments in northern Lake County, Indiana, and in Lake County, Illinois, north of Highland Park. A fringe of new housing spread more than fifteen miles west of the city, into DuPage County. Suburban growth expanded in nearly every direction possible, and new housing was built close to existing developments with little "leap-frogging" into the outer fringe.

Older suburbs such as Evergreen Park, Wilmette, Clarendon Hills, and Western Springs saw the construction of many new housing developments during the 1950s. In those older suburbs, growth tended to fill in the gaps between the linear pattern typical of the railroad-focused suburbs. The 1950s also was the decade in which

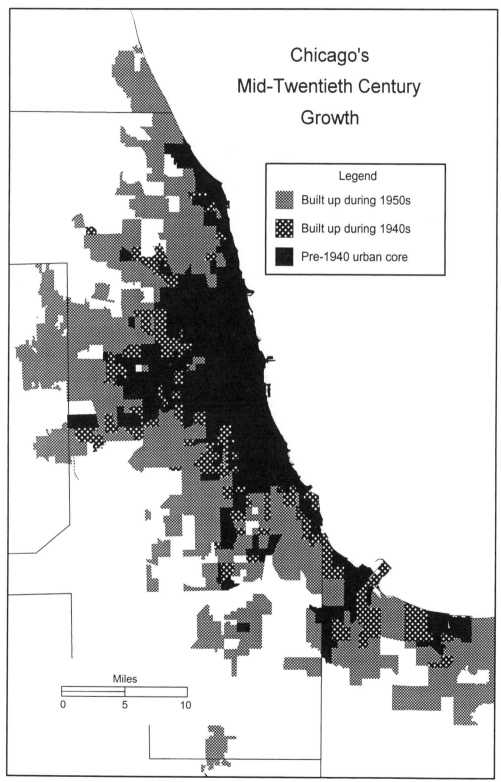

Fig. 14.2

many of Chicago's expressways were begun. The new highways extended and broadened the zone of commuter suburbs west of the city.

The 1950s would be the last decade in which Chicago's suburban growth tended to be geographically even, on all sides of the metropolitan area. Industrial growth slowed substantially in northwest Indiana after this time, and the demand for new homes in that area was less than before. The Interstate highway system and O'Hare Airport had become the new central foci of the Chicago region. Despite being crisscrossed by Interstate highways, northwest Indiana became peripheral to the new patterns of industrial growth as more and more firms chose to locate on the Northwest Side of the metropolis.

Discontinuous, or "leapfrog," suburban expansion is first evident in the 1960s (Fig. 14.3). Two distinct trends characterized the period. One was the tendency for builders to develop new housing in nearly every available tract of land within the old suburban pattern. New home developments blanketed the north suburban rail commuter line, for example, from Highland Park north to the Wisconsin border. During the same decade, the second tier of North Shore suburbs, such as Glenview, Deerfield, Northbrook, and Libertyville (one tier west of the lakefront communities), were also filled by new tract developments, usually single-family homes. A similar infilling took place along rail commuter lines and highways leading west to Elgin, Aurora, and Joliet.

Some developments in the 1960s were constructed five or ten miles beyond the continuously built-up zone of the metropolitan area. Some were in amenity locations, around lakes or on a stream. Others were placed farther out for the sake of anticipating the infilling that would follow. Downers Grove, Niles, Steger, and South Holland in Illinois, and Merrillville and Munster in Indiana, have housing construction dates that cluster in the 1960s. Tract homes in those places typically were surrounded for a time by stretches of farmland separating them from longer-settled areas.

SUBURBIA OF THE 1960s AND 1970s

An outward, incremental growth process such as suburban expansion is constrained by some geometric facts. At any given time, the zone of new development is akin to the circumference of a circle, which, in turn, is directly proportional to its radius. The amount of area available for new building, however, is proportional to area—the second power of radius. The amount of new housing construction must grow at an increasing rate in order to use up the total available space with each increment.

The population of the Chicago metropolitan area certainly was growing during the 1950s, '60s, and '70s, but it was not growing at an increasing rate. The new suburban housing developments mirrored to some extent the abandonment of the inner

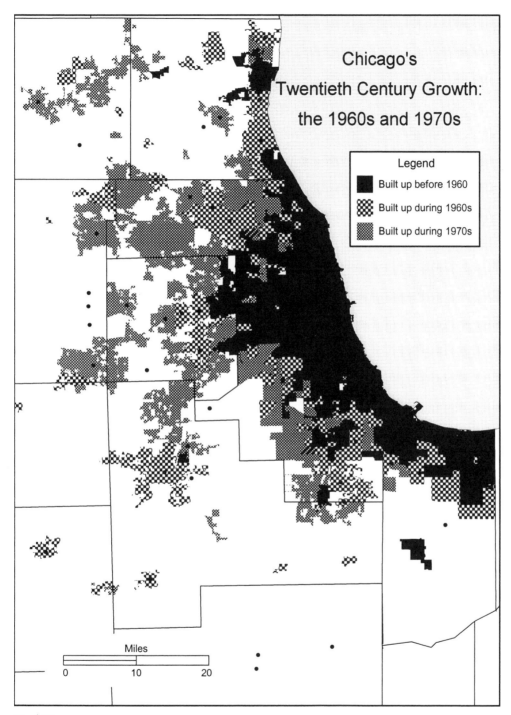

Chicago's
Twentieth Century Growth:
the 1960s and 1970s

Legend

■ Built up before 1960

▨ Built up during 1960s

▨ Built up during 1970s

Miles

0   10   20

Fig. 14.3

city by whites and their movement outward in search of more attractive settings. Such growth added to the total settled area without adding as much to the overall population. Most new growth in Chicago's suburbs from the 1950s onward took place in the suburbs, which increasingly became the destinations chosen by migrants from outside the region who "moved to Chicago." The rest of the growth was simply the other side of the white flight taking place out of the central city.

Leap-frog development was even more prevalent in the 1970s. New housing subdivisions scattered into the countryside, especially along the axes defined by Interstate highways. The urban frontier reached McHenry County in the 1970s in a chain of new housing developments constructed fifteen to twenty miles beyond the generally built-up zone. In part, the scatter was propelled by another trend that had become well established by the 1970s—the suburbanization of workplaces. Many who moved forty miles out of the city did not commute to the central city at all, but rather were employed in office parks or manufacturing industries that also had moved to the suburbs. People found they had to move outward to keep pace with the location of their jobs.

The tendency for the entire metropolitan area's economic and demographic structure to gravitate in a northwesterly direction is evident in a lack of corresponding developments along the southern edge of the metropolis. Northwest Indiana's populated areas ceased to expand outward in the 1970s. Areas that were directly adjacent—or even in the path of—expanding African-American populations in the south suburbs also grew either very slowly or not at all.

SUBURBIA OF THE 1980s AND 1990s

Townhouses, apartment-court developments, and other forms of semi-detached housing became the norm in new housing developments beginning in the mid-1970s. They were expedients to reduce the rate of sprawl, hold down infrastructure costs, and generally confine urban development in a manner that consumed less land. But they had little effect on the growth pattern of Chicago's suburbs (Fig. 14.4). Buffalo Grove, Burr Ridge, and Mettawa were new communities launched during the 1980s while Cary, Crystal Lake, Frankfort—all older towns—received substantial new growth during that decade.

Growth during the 1980s was a loose infilling of the existing pattern, spread over nearly twice as much area, given the population, as previous suburban expansions had been. The continued concentration of new-job growth northwest of the city made that sector even more attractive as a residential location.

The paucity of corresponding developments on the South Side of Chicago and its suburban fringe is equally evident in the continued slowing of growth in that direction. The cornfields started on the south side of Park Forest in 1950; they had been pushed back only a few more miles to the south by the 1990s. In the meantime, Lake

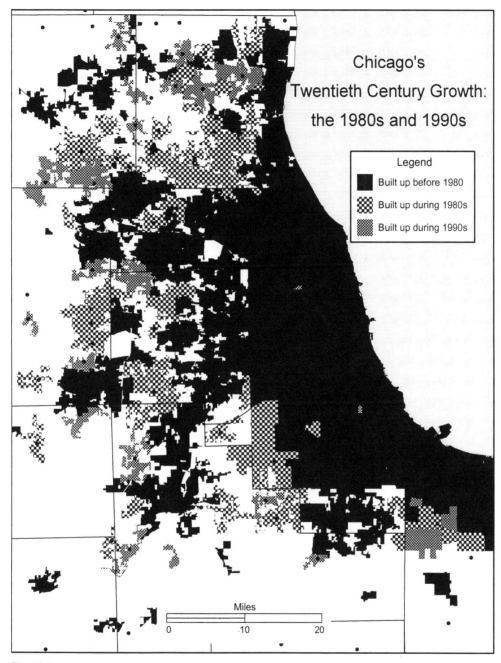

Fig. 14.4

and DuPage counties had become almost totally urbanized as had the eastern fringes of both Kane and McHenry countries. Suburban expansion slowed during the 1990s, with fewer new areas receiving developments, but the overall direction of expansion did not change.

SUBURBAN EXPANSION BY SECTORS

These trends can be quantified and summarized by changing the scale of focus from the census tract level to the village or city level (Fig. 14.5). If the metropolitan area is divided into six sectors, as shown, then the timing of new construction can be compared between sectors (Table 14.1). The southeast sector, which is mainly the Indiana suburbs, shows the least new development, especially after the 1960s. New growth was nearly evenly divided between the north, northwest, and west sectors versus the southwest, south, and southeast sectors during the 1960s; but, during the 1970s, more than two-thirds of the new communities were launched to the west and north of the city. Growth of new housing also was comparatively less significant in Table 14.1

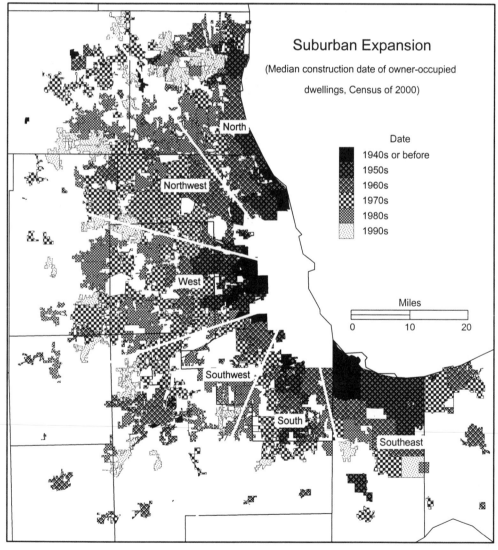

Fig. 14.5

*Median construction Date of Owner-occupied
Dwellings, 325 SuburbanCommunities, by Decade*

SECTOR:	1940S OR BEFORE	1950S	1960S	1970S	1980S	1990S	TOTAL
North	5	12	20	10	11	9	67
Northwest	3	3	11	17	18	7	59
West	10	23	13	21	11	5	83
Southwest	3	1	14	12	11	3	44
South	1	13	16	9	2	1	42
Southeast	3	6	11	6	2	2	30
TOTALS	25	58	85	75	55	27	325

Source: U.S. Census of Population, 2000

the north sector, but there the factors also include a large amount of housing that predates the 1940s. The west sector is not only larger in area, but also far larger in the number of communities that experienced housing booms in the 1970s and most other decades.

Areas that lay in the expansion path of African-American populations on the southern fringes of Chicago saw many fewer developments by the late decades of the twentieth century. This fact, which was really another manifestation of white flight, was only reinforced by the net shift in workplaces toward the northwest. The unbalancing of Chicago's longstanding symmetric structure of economy and residence was a product of both racial avoidance and economic restructuring.

Another dimension of suburban expansion was the development of outlying shopping centers to serve a more dispersed population (Fig. 14.6). More than a dozen planned, regional shopping centers were constructed in the Chicago region. Department store chains, some new to the Chicago area, prevented a further slide of retailing activity to the suburbs by building the "Magnificent Mile" development of North Michigan Avenue as an attractive shopping area during the 1970s.

The map of per capita retail sales in 2002 shows that downtown Chicago remains a very important retail center, although half a dozen outlying shopping centers are its strong competitors. Some parts of the inner city have low values of per capita retail trade because they are low-income areas. Low per-capita retail sales values in the inner city also reflect the fact that retailers have moved on and that there are fewer places to shop for inner-city residents of any income level. Many who live in the inner city must drive to the suburbs to satisfy shopping needs.

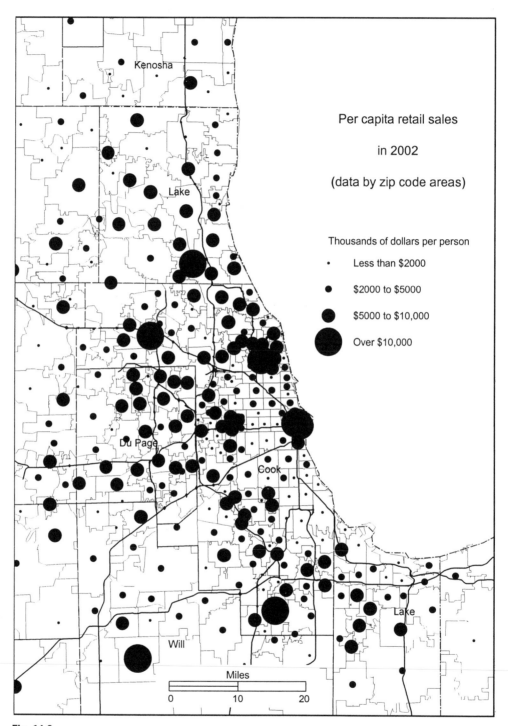

Per capita retail sales

in 2002

(data by zip code areas)

Thousands of dollars per person
· Less than $2000
● $2000 to $5000
● $5000 to $10,000
● Over $10,000

Miles

0   10   20

Fig. 14.6

Chapter 15
CHANGING ETHNIC PATTERNS

CHICAGO HAS BEEN A CITY OF IMMIGRANTS ever since its earliest years. In 1910, during the period of heaviest trans-Atlantic migration, more than one-third of Chicago's inhabitants were European-born. Another forty percent had at least one parent born in Europe, leaving a minority of only twenty-two percent of residents who had both parents born in the United States. A reduction in European immigration and the arrival of African-Americans in the city decreased the European-born component to roughly one-fifth of the city's population by 1940. But those trends, too, were later overshadowed by new patterns of immigration. The 2000 census enumerated 629,000 foreign-born persons in Chicago, only about 40,000 fewer than were counted in 1940. This small change in total numbers conceals a massive shift in terms of countries of origin, however, and the changes in nationality have given rise to ethnic residential patterns unknown in earlier times (Table 15.1 on page 182).

The changes can be interpreted in comparison with the traditional patterns of European ethnicity in Chicago. In 1950, nearly two dozen countries were represented with more than 5,000 immigrants each in the city. The 25,000 Czechoslovakians living in Chicago had a residential pattern typical of the smaller European groups (Fig. 15.1). Roughly half of them lived in the Little Village and Lawndale neighborhoods of the city's Far West Side. As many more lived in the suburbs of Cicero and Berwyn immediately to the west (not shown on the map). A half dozen other clusters of Czechoslovakian-born people were scattered over the city, including small areas adjacent to Calumet Harbor's zone of heavy industries, in the Back of the Yards neighborhood on the Near Southwest Side, and in the Pilsen neighborhood nearer the city's center. Czechoslovakians rarely formed a majority of the population except for small areas within the neighborhoods where they resided, but most of

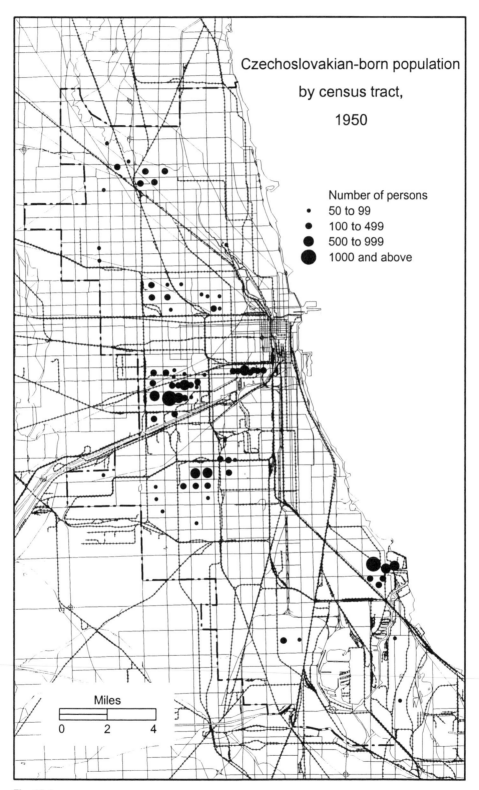

Czechoslovakian-born population
by census tract,
1950

Number of persons
· 50 to 99
● 100 to 499
● 500 to 999
● 1000 and above

Miles
0   2   4

Fig. 15.1

them lived in tracts with at least several hundred others of the same national origin within a five- to ten-block radius.

Much the same can be said about the pattern of Lithuanians living in Chicago in 1950 (Fig. 15.2). Almost identical with Czechoslovakians in terms of total numbers, the Lithuanian-born resided in Pilsen, Bridgeport, Back of the Yards, and Pullman—all adjacent to established industrial districts. The largest Lithuanian concentration was in Marquette Park, a Southwest Side residential area that grew rapidly from the influx of various immigrant groups during the 1920s. Each of the neighborhood clusters supported Lithuanian cultural institutions, churches, and social clubs. Like other European ethnic groups, very few Lithuanians lived near African-American residential areas.

The Swedish-born offer a contrasting example from 1950 (Fig. 15.3). Swedes once had been among Chicago's largest national groups, but their numbers dwindled due to a lack of replacement through immigration in succeeding generations. Death and outmigration had reduced the city's Swedish population to 46,000 in 1940, which was further reduced to 31,000 in 1950, the date of the map. (Only 563 Swedish-born people resided in Chicago in 2000.) The Andersonville community on the North Side remained as the largest concentration in 1950, although persons born in Sweden lived in many other parts of the city as well, including neighborhoods on the West, Northwest, and South sides of the city. Typical of the older European groups, Swedes had moved out from their earliest enclaves and had a more dispersed pattern of residence.

HISPANIC POPULATIONS

Only 9,000 Mexicans lived in Chicago in 1950 (Fig. 15.4). At that time, they were outnumbered by Greeks, Yugoslavians, Norwegians, Hungarians, and about a dozen other European nationalities. The Mexican-born were concentrated in the densely settled immigrant ghetto on the Near West Side. They also lived in blue-collar neighborhoods near the Calumet industrial zone and near the Union Stock Yards. Immigrants from Mexico were slightly more willing to live on the fringes of African-American neighborhoods than were any of the European-derived groups, although the residential pattern Mexicans would evolve would be distinctively their own. Between 1950 and 2000, Chicago's Mexican-born population grew from 9,000 to 292,000, while persons of Hispanic or Latino ancestry came to account for more than one-fourth of the city's total population.

The dynamics of Hispanic population growth during the latter decades of the twentieth century was enmeshed with the other major demographic trends of that era: expansion of the area occupied by African-Americans, abandonment of most parts of the central city by European-derived peoples, and a rapid growth in suburban populations. Slum clearance during the 1960s forced Mexican-Americans out of

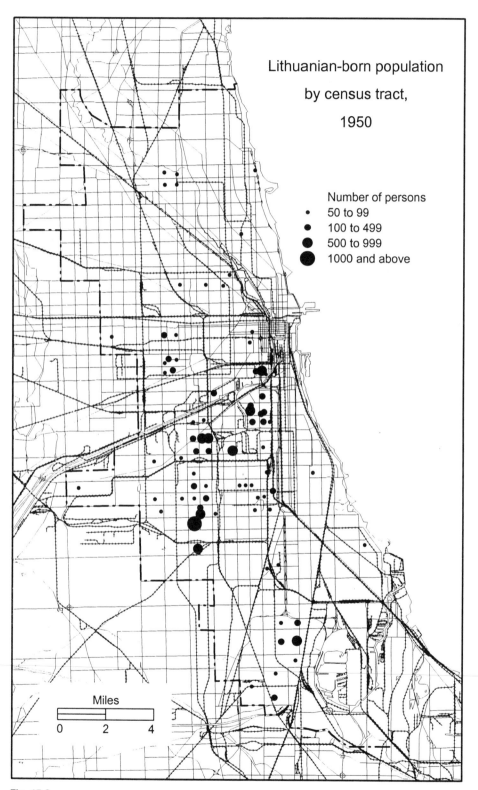

Number of persons
- 50 to 99
- 100 to 499
- 500 to 999
- 1000 and above

Lithuanian-born population by census tract, 1950

Miles

0   2   4

Fig. 15.2

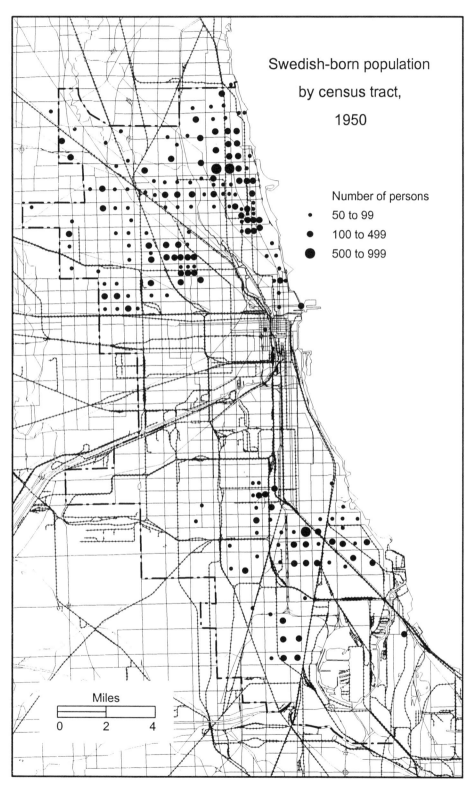

Swedish-born population
by census tract,
1950

Number of persons
· 50 to 99
● 100 to 499
● 500 to 999

Miles
0   2   4

Fig. 15.3

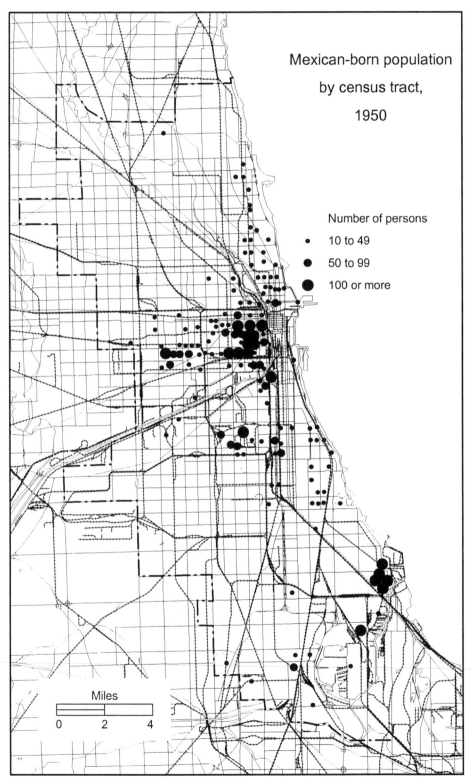

Fig. 15.4

their earliest inner-city neighborhoods. By 1970, Hispanic populations had moved both north and south to establish new residential areas in the Humboldt Park, West Town, and Lower West Side neighborhoods (Fig. 15.5). The early concentration near Calumet Harbor also expanded, but less rapidly than did the new Hispanic enclaves on the West Side.

Further growth of these three distinct areas of Hispanic residence took place during the 1980s and 1990s. Again the expansion was directed outward, toward areas of newer housing and away from the center of the city. A slender but significant axis of Hispanic population growth also appeared along the CTA Howard Street rapid transit line on the Far North Side. Such an extension, which was no doubt defined by commuting access to jobs in the city center, is an old pattern that many urban ethnic groups displayed at some time in their history.

Mexican-born people make up more than seventy percent of Chicago's Hispanic population today, as has been true in recent decades. Puerto Ricans are next in importance, with about fifteen percent of the Hispanic total. Puerto Ricans moved into many Mexican neighborhoods on the city's North Side during the 1960s. By the end of the twentieth century, the Puerto Rican population, still growing substantially, was migrating north and west more rapidly than any other Hispanic group. (A "significant" Puerto Rican population, as shown in Fig. 15.5, is any census tract with more than 250 Puerto Rican-born inhabitants.) By 2000, the expanding Puerto Rican neighborhood was advancing into areas that had been largely Polish for several decades, making the Northwest Side of the city one of rapid turnover in housing and population.

Although the movement of any "new" ethnic group into a neighborhood previously occupied by another necessarily produces a housing turnover, the expansion of Mexican neighborhoods rarely gave rise to the sudden, massive turnovers that had greeted African-Americans a decade or two earlier. Within Chicago, only twenty-five census tracts, most of which are in the Little Village and Pilsen neighborhoods, had populations that were over ninety percent Hispanic in 2000. For African-Americans, the corresponding total was 308 census tracts with an African-American population exceeding ninety percent of the total. Most Hispanics live in mixed neighborhoods that range from twenty-five to seventy percent Hispanic, with the remaining population made up by many other national groups.

The most dramatic divergence between the residential patterns of Mexicans and African-Americans comes from outside the city limits (Fig. 15.6). Some 138,000 Mexican-born persons reside in Cook County outside the Chicago city limits, and another 75,000 live in suburban DuPage and Lake counties. Taking the metropolitan area to its broadest extent—the Chicago-Gary-Kenosha CMSA—the total Mexican-

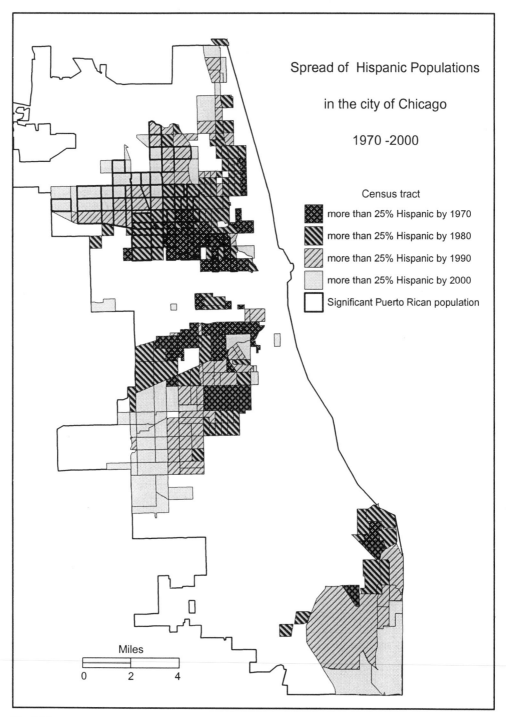

Spread of Hispanic Populations

in the city of Chicago

1970 -2000

Census tract

more than 25% Hispanic by 1970

more than 25% Hispanic by 1980

more than 25% Hispanic by 1990

more than 25% Hispanic by 2000

Significant Puerto Rican population

Miles

0   2   4

Fig. 15.5

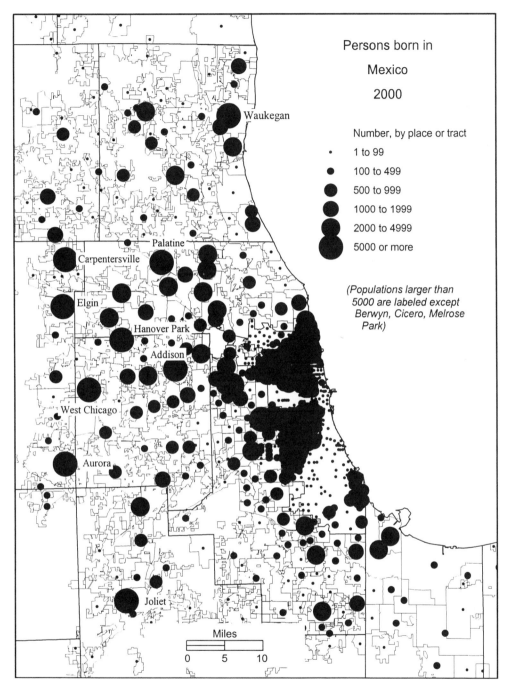

Persons born in
Mexico
2000

Number, by place or tract

·   1 to 99

●   100 to 499

●   500 to 999

●   1000 to 1999

●   2000 to 4999

●   5000 or more

*(Populations larger than
5000 are labeled except
Berwyn, Cicero, Melrose
Park)*

Waukegan

Palatine

Carpentersville

Elgin

Hanover Park

Addison

West Chicago

Aurora

Joliet

Miles

0   5   10

Fig. 15.6

born population exceeds 600,000, which is almost evenly split between the city of Chicago and the total of the outlying areas.

The growth of Mexican populations in the suburbs is a product of many trends, including the broadening variety of occupations in which Mexican immigrants are employed. Mexicans are one of the few foreign-born groups that live in rural areas, where many are employed in agricultural occupations. They are well represented in the construction, transportation, and manufacturing industries of the suburbs and fill many jobs in the service and retail-trade sectors as well. The only portions of the metropolitan area that have a small Mexican-born population are the strongly African-American sections of Chicago and some of the high-income lakefront suburbs of northern Cook County.

Some outlying cities with large Mexican-born populations, including Waukegan, Elgin, Aurora, and Joliet, have fairly well-defined Mexican neighborhoods that resemble their larger-scale counterparts in portions of Chicago; but those small-city Mexican neighborhoods are not the typical surroundings for contemporary Mexican-American families. Far larger numbers reside in neighborhoods where they are part of a mixed population that also includes Asians and European-Americans. The suburban pattern for Mexicans is skewed to the north and west, as it is for most other groups, as a product of economic change taking place at a distance from the central city.

POLISH POPULATION

Polish immigrants are the last of Chicago's large European-born groups. Nearly 70,000 persons born in Poland lived in Chicago in 2000 and as many more lived in outlying cities and suburbs of the metropolitan area (Table 15.1). The number of Poles living in Chicago in 2000 represents a forty-percent reduction compared with 1940, but smaller numbers in the city are the product of Polish migration to the suburbs and not a reduction in total immigration. Easing of visa restrictions during the 1990s made it easier for Poles to come to the United States, and the decade saw heavy immigration.

In 1950, some 94,000 Polish-born people resided in three distinct wedge-like sectors of the city (Fig. 15.7). The largest concentration, on the Northwest Side, was home to more than 40,000 Polish immigrants; 13,000 lived on the West Side, and another 17,000 lived in the Southwest Side. The Northwest Side enclave originated with a small population, which was already evident by 1890, in the mixed commercial/industrial/residential zone along Milwaukee Avenue. Poles expanded incrementally outward from this original area, and by 1950 they had an established presence over most of the city's Northwest Side. Poland's comparatively large population, plus the Poles' more than century-long habit of moving from their homeland to Chicago, fed a continued expansion of all three sectors during the last half of the twentieth century.

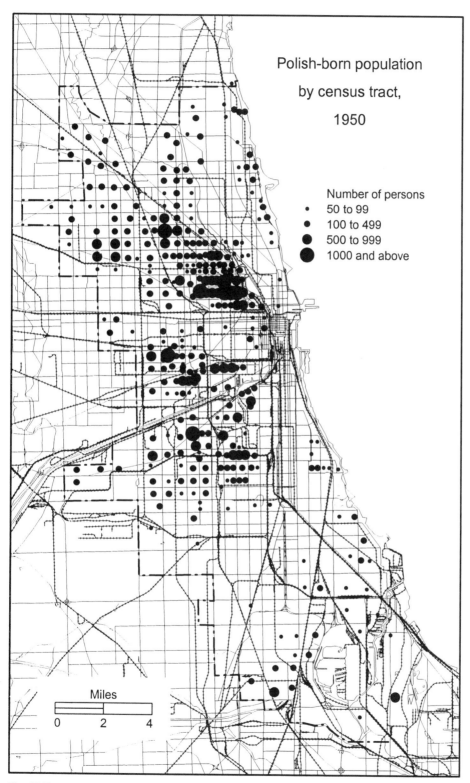

Polish-born population
by census tract,
1950

Number of persons
· 50 to 99
· 100 to 499
● 500 to 999
● 1000 and above

Miles
0 2 4

Fig. 15.7

Table 15.1 Major Sources of Foreign-born Population in the Chicago Region, 2000

NUMBER OF PERSONS BORN IN	LIVING IN		
	CHICAGO CITY	COOK COUNTY	CHICAGO-GARY-KENOSHA CMSA
Europe	145,562	288,114	379,935
Poland	69,501	120,979	139,514
Asia	112,932	231,877	333,033
China	21,725	32,179	45,832
Korea	10,167	27,397	36,468
India	15,486	48,089	78,945
The Philippines	22,678	43,564	64,555
Latin America	54,034	517,124	710,861
Mexico	292,565	430,156	602,188
TOTAL FOREIGN BORN	628,903	1,064,703	1,466,940

Source: U.S. Census of Population, 2000

Like the Mexican-born, Polish immigrants did not stop their outward expansion at the city limits. Polish people moved outward into suburbs adjacent to the city beginning in the 1950s and continued the trend in later decades. Because the timing coincided with the arrival of thousands of Hispanic immigrants to Chicago, older Polish neighborhoods became more mixed.

The outward migration continued to the point that, by 2000, the proportion of Polish immigrants had dropped below the city average in many of the tracts that had been Polish for at least half a century (Fig. 15.8). Hispanic groups typically took the place of the Poles who moved elsewhere. Although Chicago continues to have a large Polish population, the older, more densely settled inner-city neighborhoods are disappearing. In 2000, only eight census tracts of the inner city had Polish immigrant concentrations as large as five to ten percent of the total population.

The new pattern of Polish ethnicity is suburban. Their numbers are concentrated northwest of the central city, reflecting the tradition of expansion in that direction (Fig. 15.9). Seven of the eight suburban communities with Polish-born populations

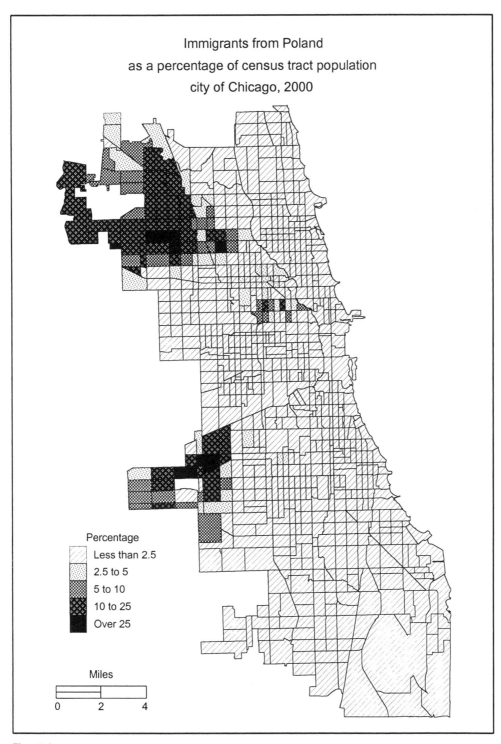

Immigrants from Poland
as a percentage of census tract population
city of Chicago, 2000

Percentage
Less than 2.5
2.5 to 5
5 to 10
10 to 25
Over 25

Miles
0   2   4

Fig. 15.8

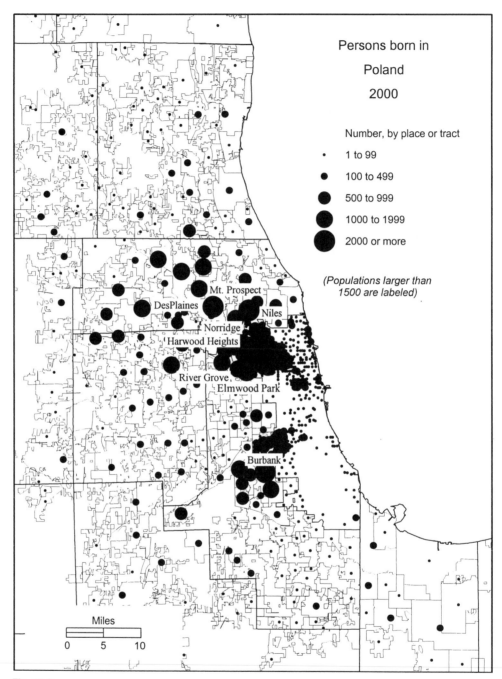

Fig. 15.9

greater than 2,000 are on the Northwest Side; the eighth, Burbank, is adjacent to the established Southwest Side Polish sector. Within Chicago, Polish-born people now live in nearly all parts of the city that are not home to substantial numbers of African-Americans. The map shows only the location of Polish-born people, not their children or grandchildren (now defined by the census as persons with "Polish ancestry"), but the same trend toward greater dispersion is even more evident for them. Like other European groups with a long history in Chicago, such as the Irish, Germans, and Swedes, Polish people have dispersed throughout the metropolitan area. Unlike those other ethnicities, however, Poles continue to favor residence in local concentrations of 1,000 or more persons of like ancestry.

ASIAN POPULATIONS

The decline of European nationalities—from ninety-five percent of Chicago's foreign-born population in 1950 to twenty-three percent in 2000—had three principal causes. One was the aging and lack of replacement of the older European migrant groups. A second was the rapid increase in migration from Mexico. The third cause was a marked increase in Asian immigration. The city of Chicago could count only 5,400 Asian immigrants in 1950. The largest concentration was of Chinese-born who lived in the Chinatown neighborhood around Cermak Road and Wentworth Avenue on the Near South Side. By the 1960s, the area was more than three-fourths Asian, mostly Chinese. Like most large-city Chinatowns, Chicago's was socially set apart from much of city life, but economically it was closely integrated with the rest of the city. Chinatown was a place of both residence and work. Chicago's Chinatown still exists today and in much the same location it has occupied for nearly a century. Only two census tracts in the city have a majority Chinese-born population, and both of them are in the heart of Chinatown.

Although many immigrants from China enter life in Chicago by at least a short-term residence in Chinatown, most do not remain there for long. The new pattern of Chinese ethnicity is suburban as well as urban, and it is characterized more by dispersion than it is by clustering. Much the same is true for persons born in Korea, India, and the Philippines (Fig. 15.10). While more than 10,000 Asian immigrants live in Chicago, they rarely form more than ten percent of a census tract's population. The largest concentration is along Devon Avenue on the Far North Side where immigrants from many Asian and Middle Eastern countries have congregated. But such neighborhoods, which are well supplied with ethnic shops, restaurants, churches, and nationality-specific social organizations, are no longer the typical settings in which Asian immigrants live.

Approximately 13,000 Chinese-born individuals reside in or near the traditional Chinatown neighborhood (Fig. 15.11). This single concentration is matched by a

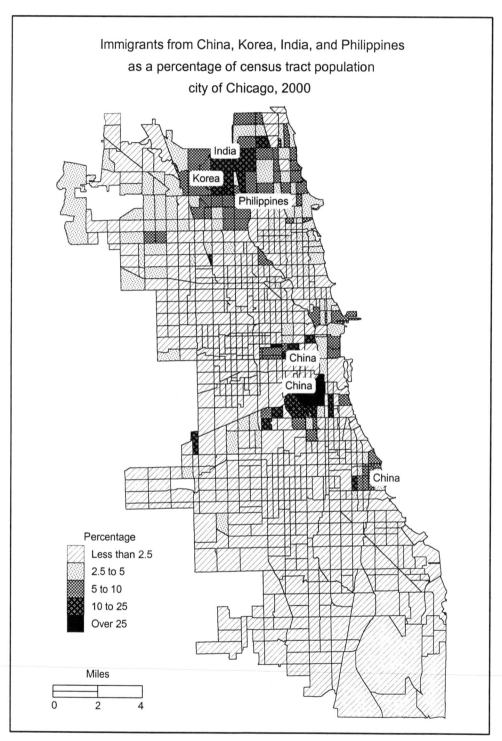

Immigrants from China, Korea, India, and Philippines
as a percentage of census tract population
city of Chicago, 2000

India

Korea

Philippines

China

China

China

Percentage

Less than 2.5

2.5 to 5

5 to 10

10 to 25

Over 25

Miles

0   2   4

Fig. 15.10

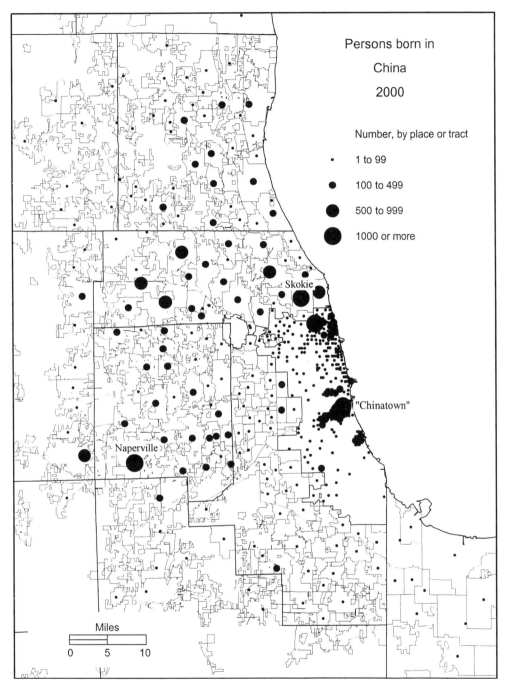

Fig. 15.11

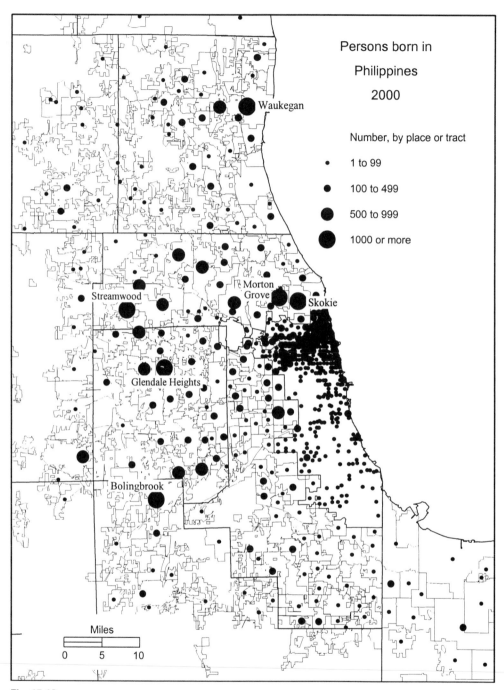

Persons born in
Philippines
2000

Number, by place or tract

- 1 to 99
- 100 to 499
- 500 to 999
- 1000 or more

Waukegan

Streamwood

Morton
Grove

Skokie

Glendale Heights

Bolingbrook

Miles

0   5   10

Fig. 15.12

roughly equal number who live in more than three dozen communities of DuPage and northern Cook counties. The Chinese, who are well represented in the white-collar professional and technical occupations, live near the places where such employment is found. Their residence pattern favors the northern and western suburbs and the high-income neighborhoods along the Chicago lakefront. Apart from a concentration in the Hyde Park/University of Chicago area of the South Side, relatively few Chinese reside south of Chinatown. Naperville's Chinese population of just over 3,000 is the largest of any suburb, but it accounts for less than three percent of Naperville's total population.

Philippine-born individuals are represented in numbers comparable with the Chinese within the city of Chicago (Fig. 15.12). Philippine immigrants have no counterpart of a Chinatown, but they are concentrated to some extent in the Far North Side tracts of the central city and in the nearby suburbs of Skokie and Morton Grove. More than 1,000 Phillipine-born people each reside in the outer suburbs of Streamwood, Glendale Heights, and Bolingbrook, all typical of the "new immigrant" pattern of the western suburbs. Because of cultural and religious affinities with Hispanic populations, Philippine immigrants also commonly live near the Mexican-born throughout the metropolitan area. Their overall pattern is scattered, although with a bias toward the northwest and a corresponding avoidance of the southern suburbs.

Koreans have been present in Chicago in significant numbers for a relatively short time. Most Koreans who have come to the Chicago area moved directly from their homeland to the Chicago suburbs. Like the Chinese, Koreans are well represented in the white-collar professional and technical occupations, and they tend to live where those types of jobs are found (Fig. 15.13). A larger number of Koreans (13,000) live in the northern Cook County suburbs than in the entire city of Chicago (10,000). Another 6,500 live in DuPage and Lake counties. Among the older suburbs, Skokie, Glenview, Northbrook, and Mount Prospect have Korean populations of more than 1,000; the latter-day communities of Schaumburg and Hoffman Estates have similar concentrations. Koreans make up a small fraction of the population in these communities, however; and, like the other Asian groups, it is difficult to name any "Korean suburbs." Koreans avoid both the South Side of Chicago and its southern suburbs.

Immigrants from India offer perhaps the best illustration of the new ethnic geography (Fig. 15.14). Few Indians lived in Chicago until comparatively recent times. Their pattern of residence in the metropolitan area reflects both their recent arrival and their proportionately large representation in the scientific, technical, and professional occupations. More than 40,000 immigrants from India reside in DuPage and northern Cook county, compared with a total of fewer than 16,000 in the city of Chicago. Seventeen suburbs are home to more than 1,000 Indian immigrants each,

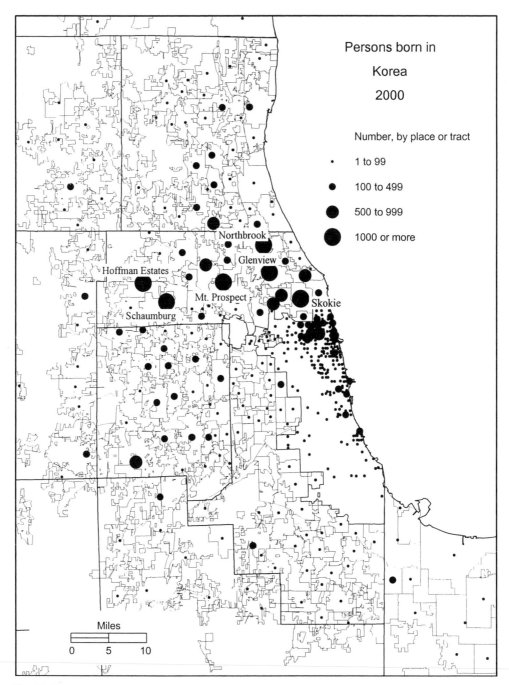

Persons born in
Korea
2000

Number, by place or tract

· 1 to 99

● 100 to 499

● 500 to 999

● 1000 or more

Northbrook

Glenview

Hoffman Estates

Mt. Prospect

Skokie

Schaumburg

Miles

0   5   10

Fig. 15.13

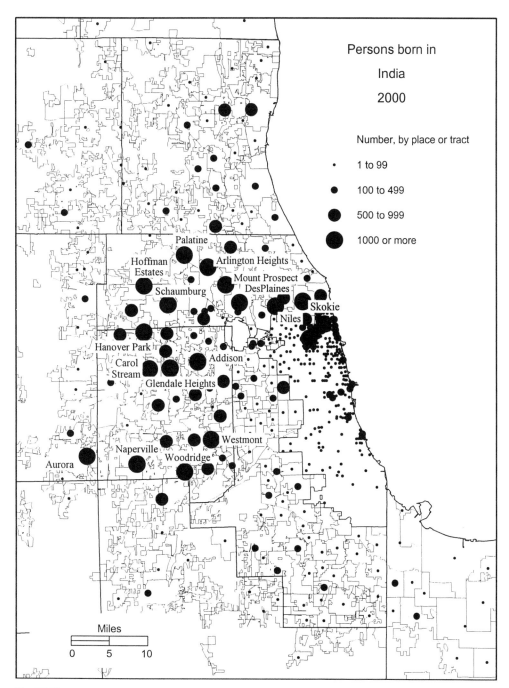

Persons born in
India
2000

Number, by place or tract

· 1 to 99

● 100 to 499

● 500 to 999

● 1000 or more

Palatine

Hoffman
Estates

Arlington Heights

Mount Prospect
DesPlaines

Schaumburg

Skokie

Niles

Hanover Park

Addison

Carol
Stream

Glendale Heights

Westmont

Naperville

Woodridge

Aurora

Miles

0   5   10

Fig. 15.14

191

but the largest number, 3,100 in Naperville, accounts for less than three percent of that city's total population.

The Japanese rank seventh among Asian immigrant populations living in the Chicago region. Although they are not as numerous, Japanese-affiliated businesses are one of the largest employers among foreign-based firms in the area. All of the more than 400 Japanese-affiliated companies in the region are located either in the city of Chicago or in the northern and northwestern suburbs (Fig. 15.15). The trend toward suburban location has led a number of these firms to move from Chicago to a northwest suburban location in recent years (Taira, 2002). Japanese-affiliated firms are located in the areas where Asian populations are most prevalent. The management of these companies is dominantly under the direction of Japanese-born individuals.

Residence patterns of Chinese, Korean, Philippine, and Indian populations in the Chicago area are similar in several ways. All four groups demonstrate a greater tendency to scatter than to cluster, reversing the process that produced the old-style ethnic neighborhoods of years' past. All four are concentrated in suburban areas north and west of the central city where professional, white-collar, scientific, and technical jobs are most abundant. All four nationalities also tend to avoid residence on the South Side of Chicago or in the outlying suburbs in that direction.

LINGUISTICALLY ISOLATED HOUSEHOLDS

These four Asian populations plus the immigrants from Mexico and Poland constitute the six largest foreign-born groups in the Chicago region today. All six populations are considered to have a native language other than English, although they diverge in the extent to which they become proficient in English after their arrival.

The U.S. Census defines "linguistically isolated households" as those in which all members over age fourteen have at least some difficulty with English (Fig. 15.16). The pattern of linguistically isolated households in Chicago and the nearby suburbs is much more reflective of Mexican and Polish residence than it is of Asian, whether on a country basis or collectively. The wedge-like concentrations of linguistically isolated households radiating from the central city mirror the pattern of both Polish and Mexican populations.

The map also suggests that those who are less inclined to learn English are not confined to older, inner-city neighborhoods. The proportion of linguistically isolated households is as evident beyond the city limits as it is within Chicago. Linguistic isolation shows no sign of diminishing, as both the Polish- and Mexican-born populations continue to expand outward into the suburban zone.

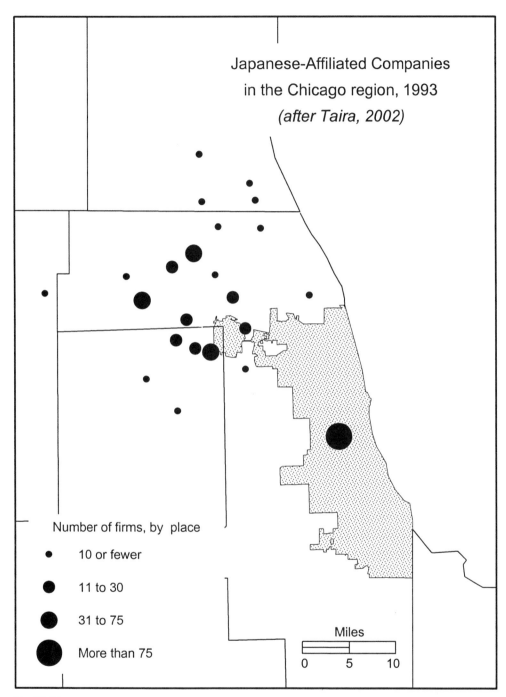

Number of firms, by place

- 10 or fewer
- 11 to 30
- 31 to 75
- More than 75

Japanese-Affiliated Companies
in the Chicago region, 1993
(after Taira, 2002)

Miles

0   5   10

Fig. 15.15

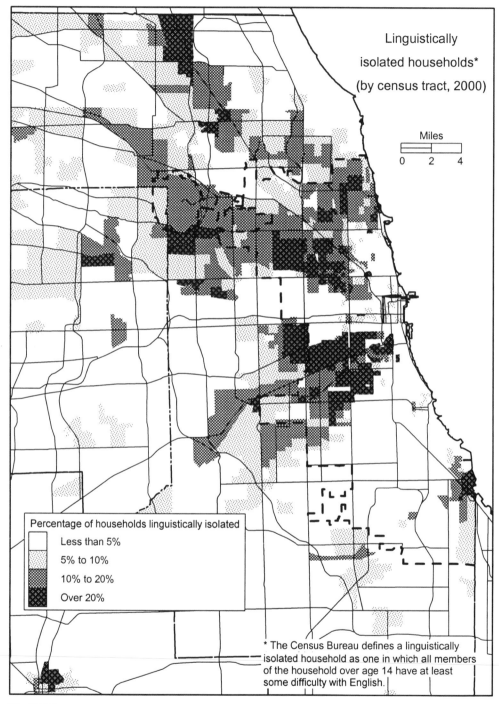

Linguistically
isolated households*
(by census tract, 2000)

Miles
0   2   4

Percentage of households linguistically isolated
Less than 5%
5% to 10%
10% to 20%
Over 20%

* The Census Bureau defines a linguistically
isolated household as one in which all members
of the household over age 14 have at least
some difficulty with English.

Fig. 15.16

Part VI

CURRENT TRENDS

Chapter 16
COMMUTING TO WORK

THE PERIMETER OF CHICAGO'S DAILY COMMUTER ZONE has changed little in nearly a century. Commuting by train from slightly north of the Wisconsin border began by 1910, although the service never has been extended farther north than that. Few people who live south of Joliet commute to Chicago, nor are there many living west of Elgin or Aurora who make a daily trip to the metropolis. These three outlying cities plus the cluster of cities around Waukegan on the north are considered satellite cities (rather than suburbs) and have long been centers of employment for local populations. Each has its own commuting field.

The U.S. Census defines multiple central cities for large metropolitan areas (Fig. 16.1). North Chicago, Evanston, Elgin, Aurora, and Joliet are considered "central cities" of the Chicago Metropolitan Statistical Area. Commuting data to those cities are combined with statistics for Chicago. (Local commuting to the outlying central cities is indicated by dashed-circle patterns in Fig. 16.1.) Many other communities have similar patterns, but they are not defined as central cities, and published data on their commuting patterns are unavailable.

Thousands of persons commute to Chicago each day from distances up to fifty miles, although about three-fourths of those who do commute to Chicago live within thirty miles of the Loop. The largest numbers of suburb-to-central-city commuters reside in places nearest to Chicago, as would be expected. The pull of city workplaces is fairly equal in all directions, with substantial numbers coming from the southern, southwestern, western, northwestern, and northern suburbs. Commuting drops rather abruptly along the southern edge of Cook County, but it continues at significant levels northward into all parts of Lake County.

Until the 1960s, it was assumed that the growth of large cities such as Chicago would lead to an expanding zone of commuting. The larger the city, the more jobs,

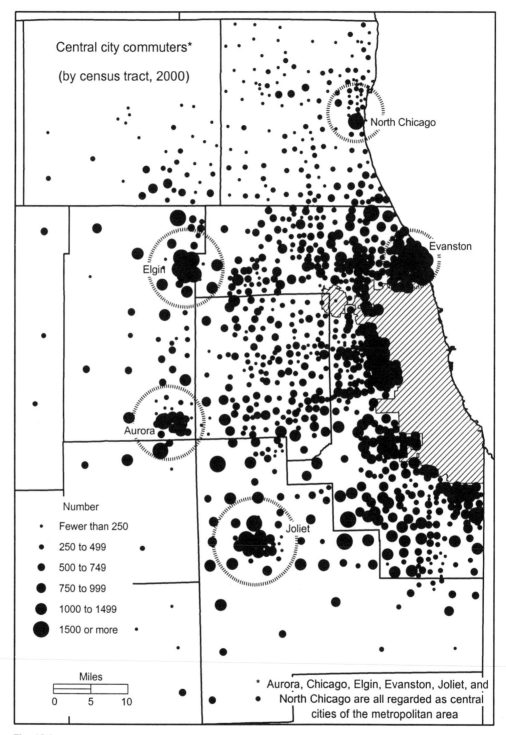

Central city commuters*
(by census tract, 2000)

North Chicago

Evanston

Elgin

Aurora

Joliet

Number

•   Fewer than 250

•   250 to 499

●   500 to 749

●   750 to 999

●   1000 to 1499

●   1500 or more

Miles

0   5   10

* Aurora, Chicago, Elgin, Evanston, Joliet, and
North Chicago are all regarded as central
cities of the metropolitan area

Fig. 16.1

and hence the greater the city's outward reach. Construction of the Interstate highway system and the general upgrading of other highways in outlying areas of the metropolis briefly had the effect of drawing more workers to the central city; but, once built, a highway running from point A to point B can be traversed just as easily in the other direction, from B to A. Rather than modernize their plants in the central city, many firms found it advantageous to move their operations into new facilities in the suburbs. Workplaces began to suburbanize in the 1960s, which gave rise to a growing trend toward reverse commuting. New businesses, with no prior history in Chicago, established workplaces in the suburbs from the start.

REVERSE COMMUTING

Thousands of Chicago residents now commute to work outside the city each day (Fig. 16.2). The tendency for them to work in the suburbs is most evident near the city's outer edges. All other things equal, a ten-mile trip to work is more likely to terminate outside the city limits for people who live near those limits. The tendency is evident over much of Chicago, although reverse commuting is much less evident on the South Side than it is on the North Side.

Reverse commuting is most prevalent in two areas of Chicago that are not close to its perimeter. One is the lakefront, stretching the entire distance from downtown Chicago to Evanston. The other is the Northwest Side. Both of these areas have a high percentage of employment in professional and other white-collar occupations. Both areas also embrace most of the neighborhoods with fairly sizable Asian populations. At the other extreme, portions of Chicago having the largest African-American populations show the least tendency for reverse commuting.

Professional, scientific, and technical services firms contribute many of the jobs that have stimulated reverse commuting (Fig. 16.3). The greatest concentration of these employers is in or near downtown Chicago. Apart from downtown, such firms are widespread over DuPage and northern Cook counties. Lake County, eastern Kane County, and southeastern McHenry County have many firms in this sector as well. Professional, scientific, and technical services firms are notably absent from much of the South Side, and the numbers are not much larger in the southern suburbs. Merrillville, Indiana, is a lone outlier on the southeast.

While reverse commuting involves thousands of workers in perhaps hundreds of separate economic activities, there is an overall northwestward bias in the distribution of both commuters and the jobs to which they travel. Employment in professional, scientific, and technical services occupations probably is as desirable as any category that could be named. It is a growing sector that represents a strain of vitality in the overall metropolitan economy. Immigrants from Asia have a residential pattern that strongly reflects the availability of such jobs nearby. These are some of the

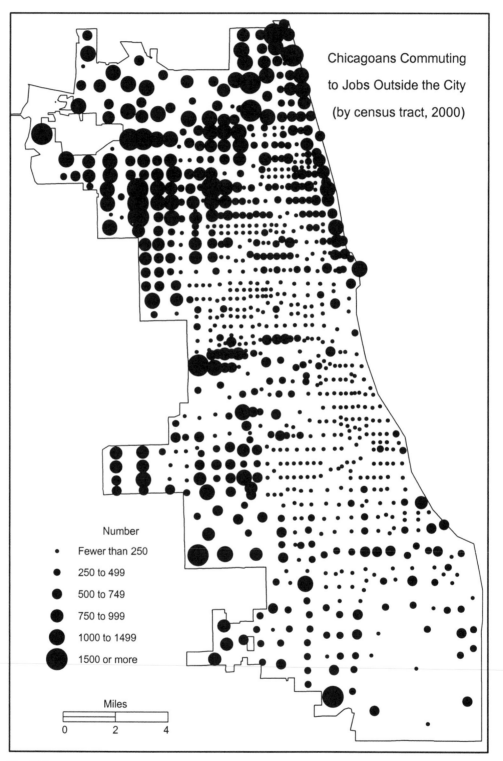

Chicagoans Commuting to Jobs Outside the City (by census tract, 2000)

Number
- Fewer than 250
- 250 to 499
- 500 to 749
- 750 to 999
- 1000 to 1499
- 1500 or more

Miles
0   2   4

Fig. 16.2

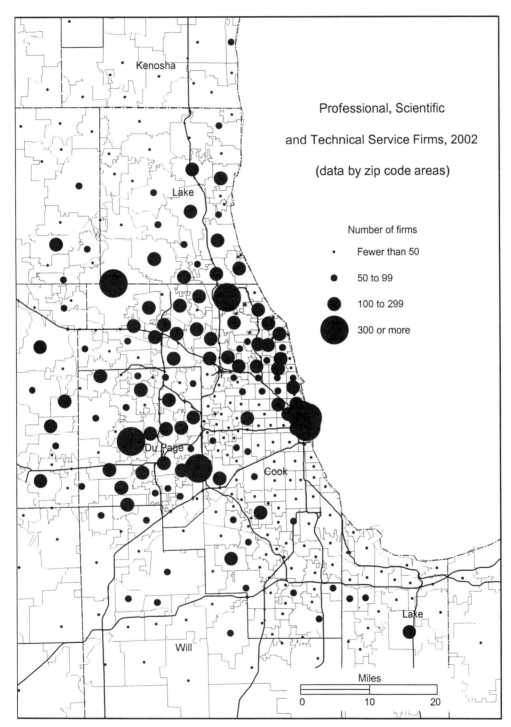

Kenosha

Lake

Professional, Scientific

and Technical Service Firms, 2002

(data by zip code areas)

Number of firms

·   Fewer than 50

●   50 to 99

⬤   100 to 299

⬤   300 or more

Du Page

Cook

Lake

Will

Miles

0   10   20

Fig. 16.3

best jobs to be had in the Chicago region, they typically demand a highly skilled workforce, and they tend to pay high wages.

There is a corresponding lack of such jobs in the southern half of the metropolitan area. On the average, residents of Chicago's South Side would have to commute approximately twice as far to such a job as would a resident of the North or Northwest sides. People on Chicago's South Side remain firmly within the orbit of central-city commuting, as they have for many years (Fig. 16.1).

The balance between central-city and suburban commuting has a break-even line near the Chicago's city limits on the west, but somewhat beyond the city limits on the south (Fig. 16.4). Gary, East Chicago, Joliet, Aurora, Elgin, Evanston, and North Chicago—all considered "central cities" of the greater metropolis—have small com-

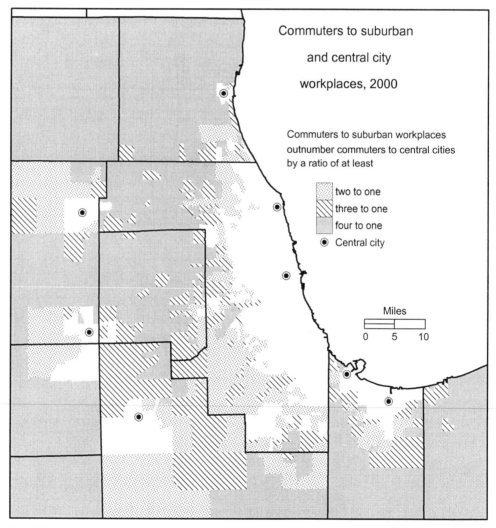

Fig. 16.4

muting fields that can be inferred from the map. The overall pattern suggests that reverse commuting probably is as common as central-city commuting for the average employed person in the Chicago metropolitan area; but the two types of commuting have very different geographical distributions.

Despite the sizable numbers of commuters from northwest Cook and DuPage counties to Chicago, most people who live in those areas now also work in the suburbs. West of the Tri-State Tollway, commuters to Chicago are outnumbered at least four-to-one by people who commute to jobs in the same or another suburb. Few portions of Kane, McHenry, or Lake counties generate many commuters to Chicago. They are places of employment in their own right, with a corresponding pattern of predominantly local trips to work.

Reverse commuting can be accomplished using public transportation, of course, but there is an inherent asymmetry that works against its success. Public transportation works best as a collector of people who live in many places but are commuting to a much smaller number of places where they work. If the employment centers also are widely scattered, as is the typical pattern in the suburbs, then automobile commuting may be the only choice.

AUTOMOBILE COMMUTING

The number of automobile commuters per household varies systematically over the metropolitan area (Fig. 16.5). Along the lakefront and within the major rapid-transit corridors, commuting by automobile is less common, probably because it is less necessary. In outlying portions of the city, in the nearer suburbs, and away from rapid transit in general, the average household has a single automobile commuter. This statistic, represented by the category .75 to 1.25 automobile commuters per household in Fig. 16.5, is a fairly widespread average. Farther out, especially in the most recently developed suburban areas, the typical household has more than 1.25 automobile commuters. In most cases, this means two working adults, one or both of whom commute by automobile, probably in separate vehicles.

Fortunately for the highway system, population density decreases outward toward the periphery at about the same rate as the tendency to commute by automobile increases. This limits commuter traffic on the highways to some extent. Reverse commuters approach suburban workplaces from the city, however, and they add congestion in places that can be difficult to predict. The construction of double or even triple left-turn lanes at what otherwise appear to be uncongested outlying road intersections has become common. They are built to accommodate traffic surges at certain times of the day resulting from automobile commuting to scattered workplaces.

There is no doubt that automobile commuting is most prevalent in those communities and neighborhoods where reverse commuting is the norm (Fig. 16.6). The

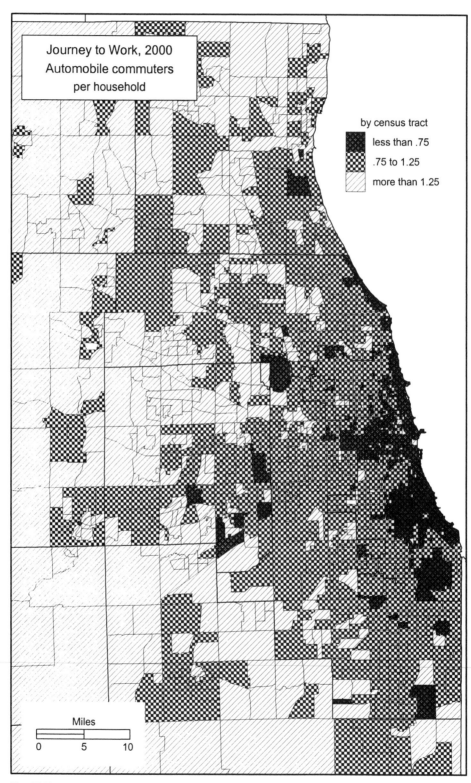

Journey to Work, 2000
Automobile commuters
per household

by census tract

less than .75

.75 to 1.25

more than 1.25

Miles

0   5   10

Fig. 16.5

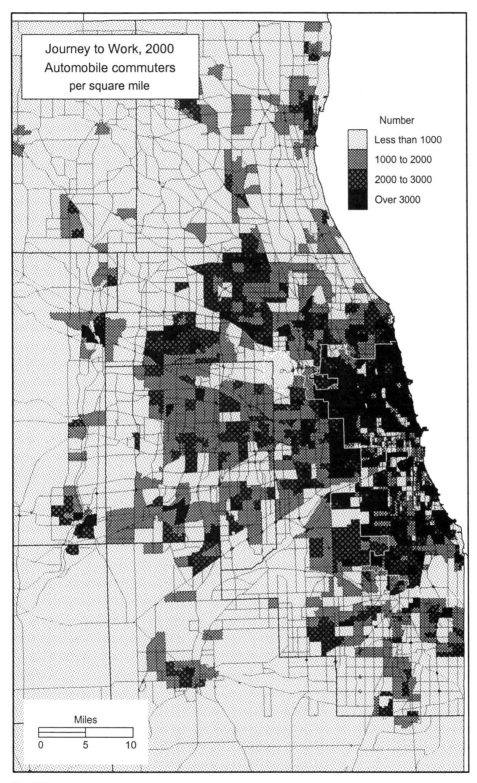

Journey to Work, 2000
Automobile commuters
per square mile

Number

Less than 1000
1000 to 2000
2000 to 3000
Over 3000

Miles

0   5   10

Fig. 16.6

association is evident on Chicago's Northwest Side, which is a densely populated area and has many reverse commuters. High-population densities and high proportions of reverse commuters produce the largest concentrations of automobile commuters per square mile in the lakefront and Northwest Side portions of the city of Chicago. Public transportation is available in those areas, of course, and it is used by persons commuting to downtown Chicago, but it is used much less often by those commuting in the opposite direction.

The map of automobile commuters per square mile can be used to gain an impression of where traffic congestion resulting from commuting would be most expected. The lack of freeways within five miles of the lakefront means that city streets on the North Side of Chicago must bear the burden of reverse commuter flows twice per day. Evanston, Wilmette, and other lakefront communities to the north have no freeways, and traffic originating in Chicago must pass through these places in order to reach suburban employment locations. Automobile commuting has no consistent pattern on Chicago's South Side where adjacent census tracts vary considerably in terms of mode of transportation to and from work.

A majority of outlying tracts that have fewer than 1,000 automobile commuters per square mile typically have values this low because of their smaller population densities; hence, they produce fewer commuters of any kind. But not all tracts at a distance from Chicago are in the lowest category. Northern and northwestern Cook County have concentrations of automobile commuters per square mile that rival many portions of the inner city. The map suggests that most of these north Cook commuters, as well as many in DuPage and Will counties, must drive at least five miles on local surface roads. This much can be inferred because most of the heavy concentrations of automobile commuters are that far from any expressway. Morning and evening traffic congestion would thus be expected in scattered, peripheral locations of the metropolitan area.

MASS TRANSIT COMMUTING

METRA is the public agency responsible for operating Chicago's commuter rail service. Those who commute to work on its trains exhibit a pattern of daily trips that once was regarded as synonymous with the journey to work in a large city (Fig. 16.7). METRA's lines have changed very little in the past seventy-five years. A new line north to the Wisconsin border, which began service in the 1990s, is the only significant addition, although minor extensions or cutbacks have been made on other lines. The equipment used in METRA service itself has been replaced no more than twice during the same span of time. Rail commuting retains the advantages it has always had; namely, that it is cheap, relatively trouble-free, and can absorb substantial fluctuations in ridership without difficulty. No better means of travel from the outlying suburbs to downtown Chicago has been devised. For those who can make use of the service, it is the best choice to be made.

But railroad transportation does not help solve most of the problems posed by reverse commuting. Trains can move large numbers of people to a single destination, but they cannot distribute those same numbers over a broad area. PACE, the public bus line of Chicago's suburbs, has introduced connecting bus services between outlying railroad stations and major employment sites. Some businesses that employ large numbers of workers at distant locations also have introduced company-supported bus services linked to railroad stations. Such services have yet to produce a significant increase in mass transit ridership among reverse commuters.

Another perspective on reverse commuting is revealed by the scarcity of METRA commuters living in the heart of Chicago. (Fig. 16.7 is based on census data that records mode of commuter travel in terms of where people live, not in terms of where they work.) Although downtown Chicago is not primarily a residential area, anyone who does live downtown is well served by the METRA network. High-speed access to outlying areas is available in no fewer than twelve radiating corridors. No place in the metropolitan area would be more advantageous for a reverse commuter to live, given that all commuter services terminate at stations no more than a few blocks apart in the center of the city; but those who live in downtown Chicago typically work there as well (Fig. 16.2). The number of METRA riders living in downtown Chicago is as small as anywhere in the city.

METRA primarily serves the economic sectors that are concentrated in the heart of Chicago. Banking, financial services, commodities trading, and other businesses that deal in monetary transactions are among the higher-income occupations found in or near the Loop. Lawyers and others dealing in legal services, public or private, are concentrated in the same areas. Firms engaged in advertising and architectural design are clustered in areas slightly north of downtown, as are many of Chicago's larger retail stores. Cultural institutions, government agencies, and some corporation headquarters are also downtown-centered. People employed in these activities live in substantial numbers along all METRA lines converging on downtown Chicago from all portions of the metropolitan area.

METRA commuters are most numerous along the former Burlington Northern rail line to Aurora. Naperville, which is Chicago's largest suburb, also is the commuter station contributing the largest number of METRA riders. Half a dozen other communities along this line have substantial numbers of rail commuters as well. The former Chicago and North Western "North" line to Waukegan ranks second. Evanston, Wilmette, Winnetka, and Highland Park all show significant commuting

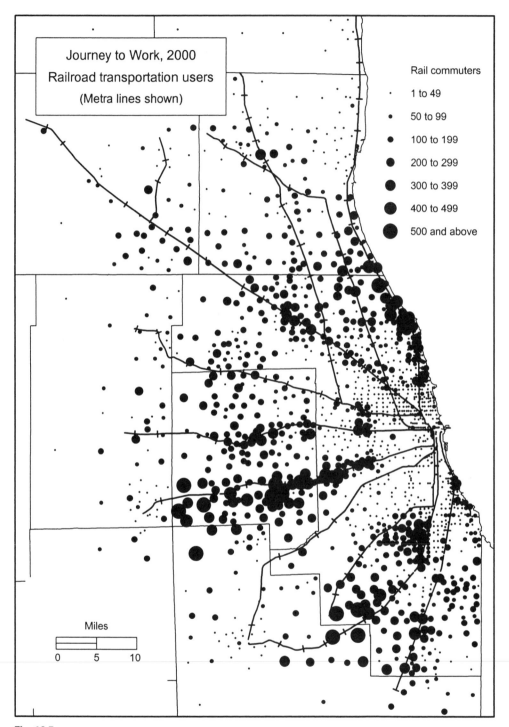

Journey to Work, 2000
Railroad transportation users
(Metra lines shown)

Rail commuters

· 1 to 49
· 50 to 99
· 100 to 199
● 200 to 299
● 300 to 399
● 400 to 499
● 500 and above

Miles
0   5   10

Fig. 16.7

by rail. A third significant corridor of rail commuting lies to the southwest of downtown Chicago. Two lines, one terminating at Joliet and the other at Orland Park, are heavily patronized. A large number of rail commuters on both of these lines make daily trips from Blue Island and nearby suburbs.

Commuting by rail decreases significantly with proximity to the Loop. Other modes of transportation, including CTA trains, take over from METRA within the city (Fig. 16.8). Unlike METRA, which attracts substantial numbers of riders on most if not all of its routes, those who commute to work via the CTA subway and elevated trains are strongly clustered around just a few routes. By far the heaviest patronage is on the line from the Loop north to Howard Street and beyond to Evanston and Wilmette. This service is heavily used by people who work in the Loop. Also heavily traveled is the CTA's elevated Ravenswood line, which diverges from the Howard Street route at Belmont Avenue.

Moderately patronized routes include the Kennedy Expressway rapid transit line to O'Hare Airport across the Northwest Side. Many commuters from Oak Park travel to downtown Chicago via the CTA's Congress Street ("Blue Line") trains, although comparatively few West Side Chicago residents patronize the same service. A new line constructed to Midway Airport in the 1990s has yet to develop a large volume of use. In general, CTA transit ridership is much heavier on the North Side of the city than it is on the South Side.

Bus routes are spaced one-half mile apart in the cardinal directions of the urban grid, on both the North and South sides of Chicago (not shown on the maps). The direction of bus ridership is difficult to determine from census data, but many routes do carry heavy volumes. Bus services are timed to coincide with arrival of CTA rapid transit trains at key service locations; hence, many commuters take both a bus and a train to work.

There are thus three well-defined patterns of commuting to work in Chicago's metropolitan area. Commuting to downtown Chicago still is the norm for thousands of people who live in all parts of the city itself and for thousands more living in the suburbs. These are the "traditional" commuters, and they are the ones who are most likely to use public transportation. While the high level of commuting to downtown by residents of Chicago's South Side and the southern suburbs demonstrates the attraction of jobs in the Loop, it also reflects a lower level of suburban employment possibilities in the southern half of the metropolitan area.

Most of DuPage and northern Cook counties and nearly all of Lake and McHenry counties are part of a different pattern. In those areas, more than four-fifths of all employed persons work somewhere outside the city of Chicago. Many of these outlying suburbs are no farther from downtown Chicago than are others that

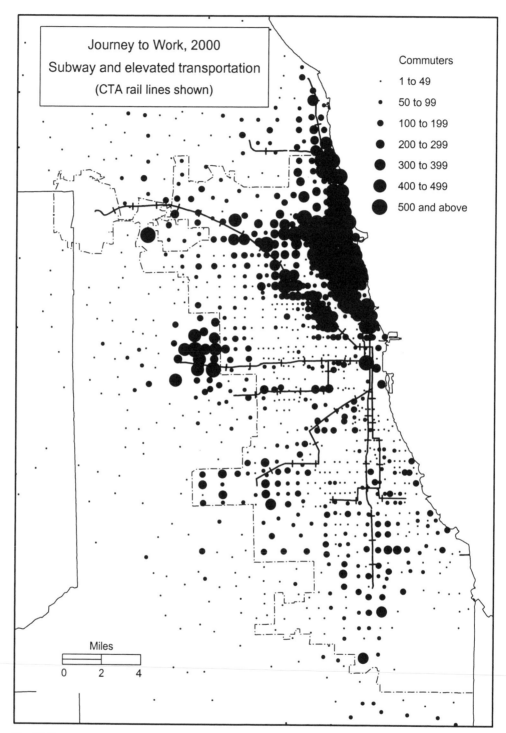

Journey to Work, 2000
Subway and elevated transportation
(CTA rail lines shown)

Commuters

· 1 to 49
· 50 to 99
● 100 to 199
● 200 to 299
● 300 to 399
● 400 to 499
● 500 and above

Miles

0   2   4

Fig. 16.8

show a higher level of commuting to the city. Many of them are well served by public transportation that enables fast commuting to downtown, although automobile commuting is the norm. These are the communities where workplaces are now becoming concentrated. Commuter trips generated within them are likely to be contained within the suburban realm.

The third group consists of reverse commuters. They are people who live either in Chicago or in one of its older, adjacent suburbs and who commute to workplaces in the more distant suburbs. Urban mass transportation is an inconvenient alternative for many reverse commuters because the workplaces to which they commute are more dispersed than are their places of residence. Traffic flows resulting from reverse commuting have placed an extra burden on local roads many miles from downtown Chicago.

Chapter 17
AGE AND NEIGHBORHOOD CHANGE

THE PASSAGE OF TIME has several impacts on the geography of a large metropolitan area. The age structure of the population measures the passage of time in generations. Even if no one in the city moved, neighborhoods would undergo a succession of changes as the inhabitants grow older. Presence of a young-adult population means the presence of children and hence a demand for schools; but, as those young parents get older and their children move on, the neighborhood is transformed. The cycle is repeated as the older generation is replaced by a younger one. At the same time, a newer, more distant suburb may be entering the early phases of growth with many new households being formed.

A second manifestation of change over time is the aging of structures. As buildings grow older they are either maintained or upgraded, allowed to deteriorate, or are removed and replaced by others. Age of the housing stock sometimes reflects the age of its inhabitants, but not necessarily. Young adults often form their first households in older housing that has been passed along from generation to generation. Structures that are allowed to deteriorate typically pass from a higher socioeconomic group to a lower one. Entirely new buildings might be occupied by people of any age. Both aging processes—of people and of structures—present a constantly changing stage on which other aspects of social change are played out.

AGE STRUCTURE
Most cities have a pattern of residence classified by age that varies systematically over the urban area. Households composed mainly of elderly persons, households with young children present, and households consisting only of young adults typically have different patterns of concentration. All three of these household types have distinct geographies in Chicago and its surrounding suburbs.

A higher than average concentration of school-age children in a neighborhood typically is associated with a predominance of householders in their thirties and forties (Fig. 17.1). While the proportion of families with children in school does not vary by more than ten percent from one census tract to another over most of Chicago, two types of areas have high values. Census tracts on the West, Southwest, and South sides of the city have larger than average numbers of children in school compared with the metropolitan area as a whole. There is no strong association with race or ethnicity, although many areas with large Hispanic or African-American populations also have larger numbers of school-age children. A second area of such concentration is found in the outer suburbs, on all sides of the city. Scattered portions of Lake, McHenry, DuPage, Kane, and Will counties, where single-family homes predominate, have higher numbers of school-age children. These are typically residential areas where most of the housing has been built within the past two decades.

Older suburbs, closer to the central city, have lower than average numbers of school-age children. Residential construction booms took place in these suburbs more than thirty years ago. The children born there have grown, passed beyond their school years, and perhaps moved elsewhere. Census tracts along the Chicago lakefront also have low numbers of school-age children. They are areas where most households are formed of young adults who have not yet begun to raise families.

To a certain extent, the map of elderly population is the reverse of that portraying school-age children (Fig. 17.2). Persons over age 65 are concentrated in the older suburbs adjacent to the central city. Many moved to those areas as young adults, when the near suburbs were growing rapidly. Older people are correspondingly fewer in number in the outer, more recently built-up urban fringe. They also are less common in those portions of the central city where school-age children are most numerous. Very few census tracts in the city of Chicago have more than one-fourth of their populations in the over-65 category. The largest concentration of census tracts with an aging population lies between ten and twenty miles from downtown Chicago, especially to the northwest of the city. These areas, which epitomized suburbia in the 1950s and 1960s, now have a large population in the older ages.

Young adults in the 20 to 29 age range have a strongly clustered residential pattern (Fig. 17.3). As would be expected, people of this age constitute a majority of residents near colleges and universities. Concentrations in Hyde Park around the University of Chicago, near the Illinois Institute of Technology campus on the Near South Side, and around the University of Illinois at Chicago campus on the Near West Side are examples.

The largest area of 20 to 29 year olds, however, is the recent college graduate residential pattern of the North Side. Between Ashland and Halsted streets and between Fullerton Avenue and Irving Park Road, the proportion of the population in the 20

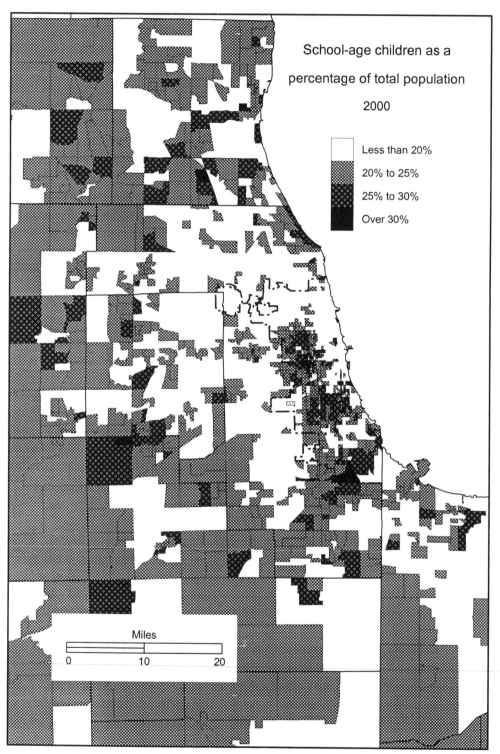

School-age children as a
percentage of total population
2000

Less than 20%
20% to 25%
25% to 30%
Over 30%

Miles

0   10   20

Fig. 17.1

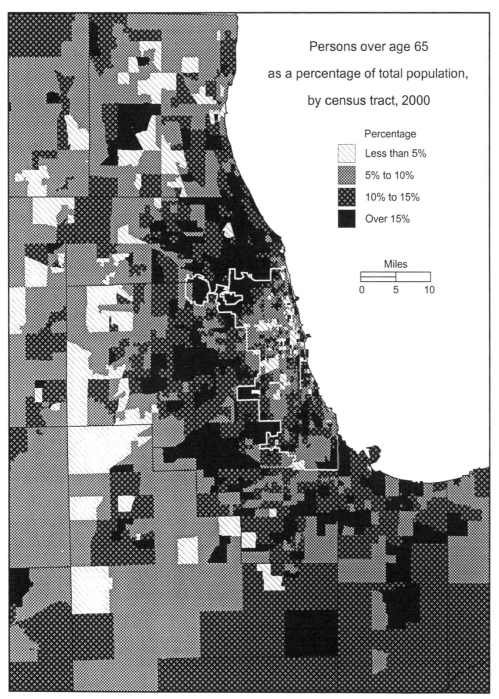

Persons over age 65
as a percentage of total population,
by census tract, 2000

Percentage
Less than 5%
5% to 10%
10% to 15%
Over 15%

Miles
0   5   10

Fig. 17.2

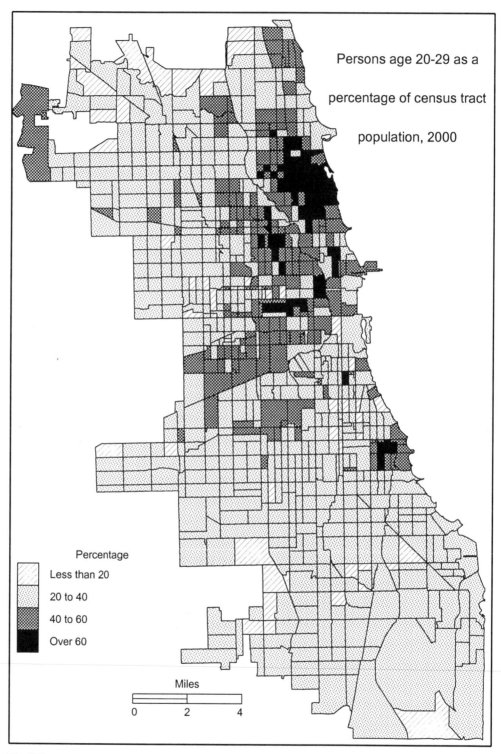

Persons age 20-29 as a percentage of census tract population, 2000

Percentage
Less than 20
20 to 40
40 to 60
Over 60

Miles
0   2   4

Fig. 17.3

to 29 category averages more than sixty percent. This area, which overlaps the Wrigleyville, DePaul, and Lincoln Park neighborhoods, is one of the most vivid examples of voluntary segregation in Chicago. It is the neighborhood of choice for many college graduates who enter the labor force in white-collar occupations. People in this age range, plus those a decade older, account for a substantial fraction of those employed in professional occupations and who live along or near the lakefront north of downtown Chicago.

NEIGHBORHOOD CHANGE

The movement of post-college young adults into Chicago's North Side neighborhoods is one facet of a larger trend toward "moving back to the city" that has gained in popularity during the past three decades. The now-common practice of housing renovation and upgrading known as gentrification is another aspect of the same trend, usually involving a slightly older and more established population. Many of those who move "back" to the city never lived there before, although many of their parents and grandparents did. Their parents were part of the massive exodus to the suburbs that took place after World War II.

Because gentrification involves replacing a low-income population with one enjoying a relatively higher income, an increase in white population typically accompanies the shift (Fig. 17.4). Chicago's white population dropped by only 48,000 during the 1990s, a comparatively small loss on a population base of roughly 1.2 million, which was the smallest white population decline in many years. Although white numbers are roughly stable at present, many more neighborhoods showed a net loss of whites than recorded a net gain. Areas that experienced any white increase also had a substantial white increase because the magnitude of change is on a neighborhood-wide scale rather than on an individual household basis. Large increases in white population accompany near-total replacement or upgrading of the housing stock in neighborhoods that whites had all but deserted two, three, or even four decades ago.

Asian immigrants also are moving into the central city (Fig. 17.5). Many of them are employed in professional and other white-collar occupations and work in or near downtown Chicago. In most cases, they are moving into neighborhoods where Asians never lived before. Their preference for younger, upscale residential neighborhoods near the lakefront is as obvious as their avoidance of most areas of the South and West sides of the city.

Asian population increases in the Portage Park, Albany Park, and West Rogers Park neighborhoods follow a more mixed trend in which recent migrants from India, China, Korea, and the Philippines mingle with immigrants from many other countries. They are employed in a wide range of occupations, from blue-collar to

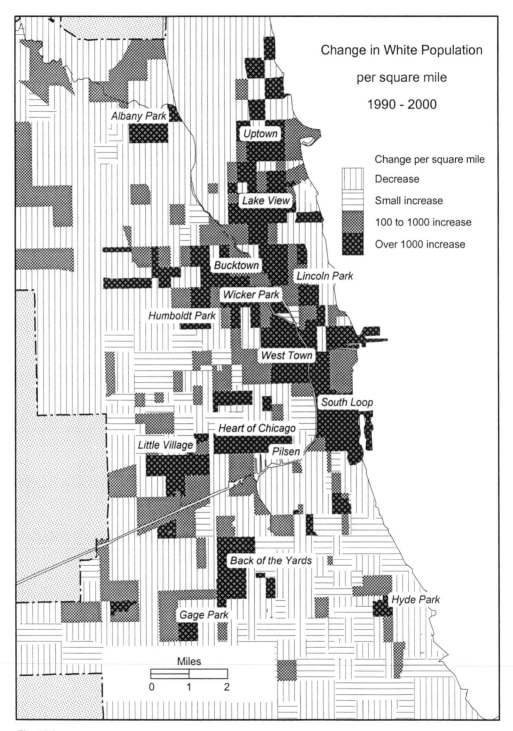

Change in White Population
per square mile
1990 - 2000

Change per square mile
Decrease
Small increase
100 to 1000 increase
Over 1000 increase

Albany Park

Uptown

Lake View

Bucktown

Lincoln Park

Wicker Park

Humboldt Park

West Town

South Loop

Heart of Chicago

Little Village

Pilsen

Back of the Yards

Hyde Park

Gage Park

Miles
0   1   2

Fig. 17.4

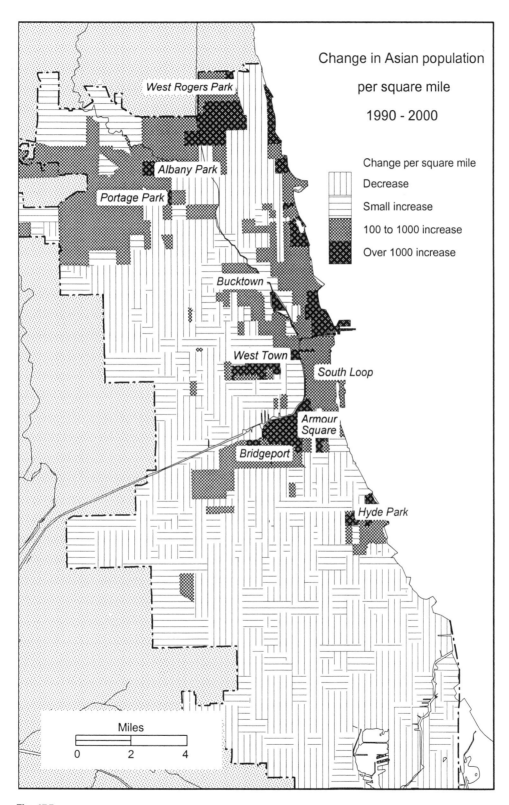

Change in Asian population
per square mile
1990 - 2000

Change per square mile
Decrease
Small increase
100 to 1000 increase
Over 1000 increase

West Rogers Park

Albany Park

Portage Park

Bucktown

West Town

South Loop

Armour
Square

Bridgeport

Hyde Park

Miles
0   2   4

Fig. 17.5

white-collar, and are less oriented to professional employment than are the recent Asian immigrants living near the lakefront.

Two kinds of neighborhood change are associated with white and Asian population increases in the central city. The more widespread trend involves a mixture of renovation and rebuilding in older, established residential neighborhoods. White and Asian population increases in the Bucktown, Wicker Park, and Little Village neighborhoods are examples of these transformations. Residence in these neighborhoods is very attractive to young singles and couples who desire proximity to downtown.

A second type of neighborhood change also involves a shift from commercial to residential land use. New housing has been created near downtown in areas that had not been residential in character for many years (Fig. 17.6). Conversion of former warehouses or factories into condominium apartments, construction of new high-rise apartment buildings, and even the appearance of new areas of detached, single-family housing have been noticeable in the inner city ever since the early 1970s. The first boom was in apartment buildings. Once the lakefront sites had been occupied by high-rise structures, attention shifted landward into former industrial areas. Although the map shows only the construction dates of owner-occupied buildings, rental properties in the same areas have similar dates of construction.

RESIDENTIAL PERSISTENCE

At the other end of the housing-age distribution are neighborhoods where older homes predominate (Fig. 17.7). Structures built before 1955 (more than forty-five years old in 2000) are most prevalent on the North and Southwest sides of Chicago. Many lakefront condominium/apartment buildings on the South Side also are of older vintage. Neighborhoods with a larger than average concentration of people over age 65 do not stand out as those where older housing predominates, however. Persons in the 20 to 29 age group are most likely to live in older housing, followed by families with children in the school-age years (Figs. 17.1-17.3). The large concentration of older housing on Chicago's Northwest Side reflects the persistence of a housing stock that has been in place for several decades longer than the maximum age shown in Fig. 17.7. One or more waves of population turnover have characterized such areas, where the actual structures have persisted while the families living in them have come and gone.

Age of structure and age of householders are both reflected in the map of residential persistence (Fig. 17.8). Neighborhoods where housing is new or where the present occupants have lived there only a short time naturally exhibit low values of residential persistence. The longest duration of residence characterizes Chicago's older African-American neighborhoods. Areas of greater than twenty year's persist-

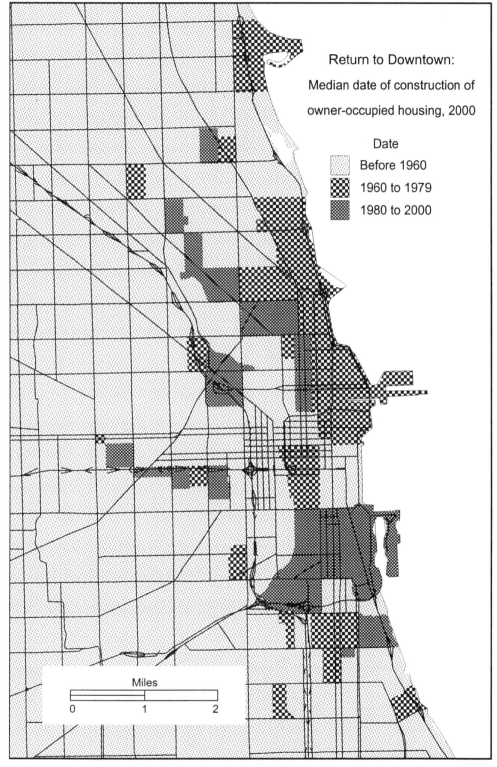

Return to Downtown:
Median date of construction of
owner-occupied housing, 2000

Date
Before 1960
1960 to 1979
1980 to 2000

Miles
0   1   2

Fig. 17.6

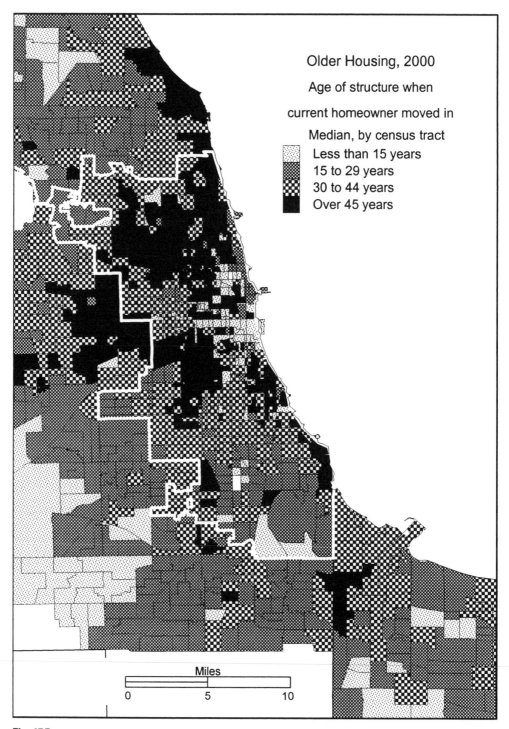

Fig. 17.7

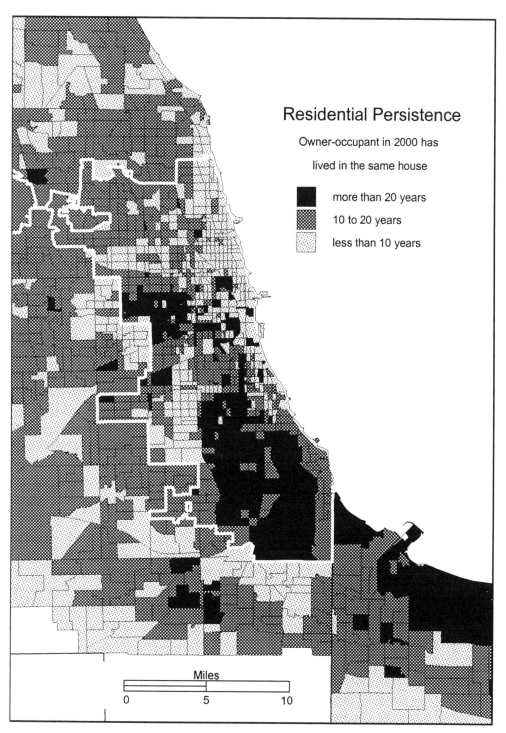

Residential Persistence

Owner-occupant in 2000 has

lived in the same house

more than 20 years

10 to 20 years

less than 10 years

Miles

0   5   10

Fig. 17.8

ence include portions of Chicago's West Side where blacks outnumber whites more than 100-to-one. On the South Side, the most stable neighborhoods in terms of low population turnover are those where blacks outnumber whites by at least fifty-to-one. Long-term residence in the same dwelling in these cases reflects the poor condition of many residential structures. The owner-occupant's investment is sunk in the dwelling. Such buildings are unsalable except at a very low price, which only hastens the process of decline. While persistence can be taken as a sign of neighborhood stability, it can also indicate stagnation and decline.

On Chicago's South Side and in adjacent portions of northwest Indiana, residential persistence characterizes white and black neighborhoods alike. In Hammond, Indiana, just across the state line from Chicago, there is a twenty-to-one ratio of whites to blacks in a large area of residential persistence. In the city of Gary, just to the east of Hammond, the ratio is reversed in an overwhelmingly black neighborhood, but one that equally demonstrates long duration of residence. For both groups, residential persistence no doubt reflects the declining economic fortunes of the local area where job losses in the heavy industries have been substantial and residential properties have not appreciated in value. The population of such neighborhoods is either growing slowly or declining.

RACE, PERSISTENCE, AND TURNOVER

A lack of residential persistence characterizes neighborhoods that have experienced a substantial recent turnover in population. Persistence values in the bottom category (less than ten years' residence) are found in scattered areas of the city that have seen a recent increase in African-American population (Fig. 17.9). The Gage Park, Chicago Lawn, and Ashburn neighborhoods on the Southwest Side are examples, as are Austin, Hermosa, and Rogers Park on the Northwest and North sides of the city. These developments represent a type of neighborhood change more reminiscent of the turbulent era of three to four decades ago. Unlike former times, however, African-Americans are now moving into many kinds of neighborhoods and are doing so in nearly all parts of Chicago.

One of the strongly negative results of African-American migration to Chicago in years past was the channeling of new migrants into the poorest, most overcrowded, deteriorating portions of the central city. This was especially true of African-Americans who, having lived in the rural South, were unprepared for the kind of life they would find in a Northern city. Thousands of African-American migrants from the Southern states still live in some of Chicago's poorest neighborhoods, but they are no longer confined to such locations (Fig. 17.10).

Newly arrived African-American migrants from the South are just as likely to take up residence in one or another of the newer African-American residential areas

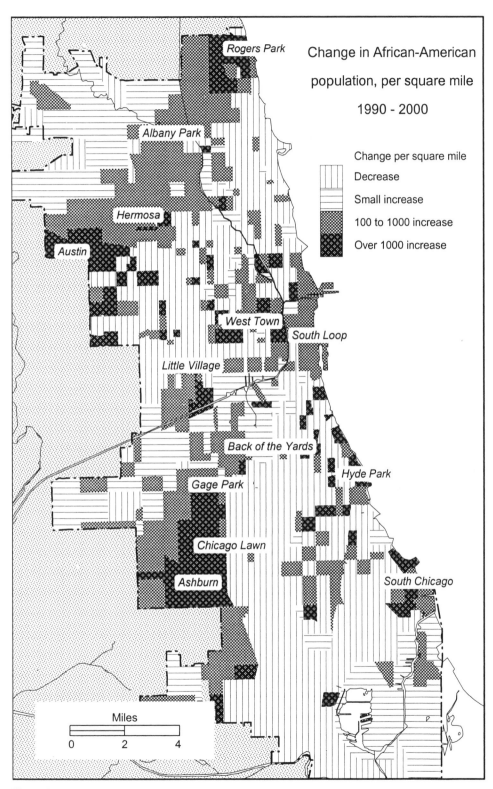

Change in African-American
population, per square mile
1990 - 2000

Change per square mile
Decrease
Small increase
100 to 1000 increase
Over 1000 increase

Rogers Park

Albany Park

Hermosa

Austin

West Town

South Loop

Little Village

Back of the Yards

Hyde Park

Gage Park

Chicago Lawn

Ashburn

South Chicago

Miles
0   2   4

Fig. 17.9

225

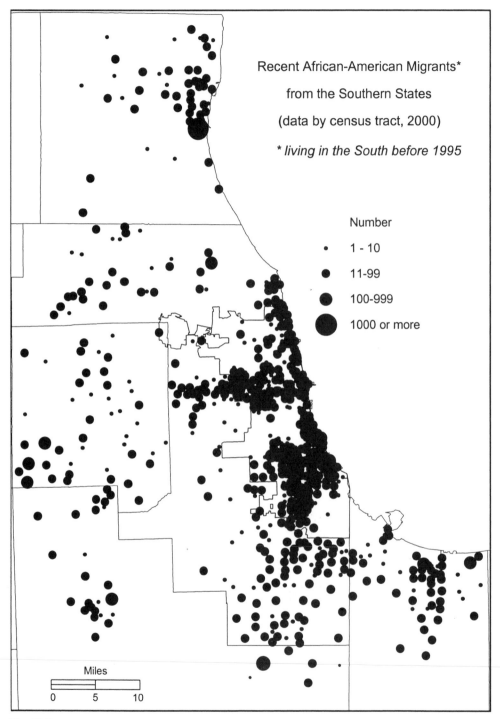

Recent African-American Migrants*

from the Southern States

(data by census tract, 2000)

living in the South before 1995

Number

· 1 - 10

● 11-99

● 100-999

● 1000 or more

Miles

0　5　10

Fig. 17.10

scattered around the suburbs as they are to locate in the central city. Southern Cook County suburbs are noticeable for the numbers of new African-American migrants they attract. Urban fringe census tracts adjoining the cities of Gary, Waukegan, and Joliet show a similar pattern. Although the map does not suggest that newly arrived African-American migrants find housing equally, in all parts of the metropolitan area it is obvious that the sharp contrasts between all-black and all-white areas that has long characterized Chicago is much less evident in the suburbs.

Chapter 18
INCOME DISTRIBUTION

Variations in household income are associated with many other aspects of social and economic life in the metropolis. The income maps presented in this chapter show concentrations at both the low and high ends of the income scale, for both the central city and the suburbs. Median household income in the Chicago metropolitan area was $51,046 in 1999, which means that exactly half the households had incomes above this value and half had incomes below it. Twenty-two percent of households had annual incomes below $25,000, and roughly seventeen percent had incomes above $100,000. While three of every five households had mid-range incomes, a focus on the upper and lower ends of the distribution reveals more about the factors that produce large differences in income distribution from place to place.

LOW-INCOME NEIGHBORHOODS

The concentration of Chicago's lowest-income households in the predominantly African-American neighborhoods of the city is now a pattern of nearly a century's duration (Fig. 18.1). Most census tracts, in which at least three-fourths of the households have incomes below $25,000, include one or more public housing projects, which almost by definition places such tracts near the bottom of the income scale. The fact that these clusters of extreme poverty are scattered over the inner city, in neighborhoods such as Cabrini Green, Wentworth Gardens, North Lawndale, and LeClaire Courts, is the direct result of the choices made by the Chicago Housing Authority in choosing sites for public housing.

But low-income households are by no means confined just to areas where the very poorest, most dependent people reside. In the majority of census tracts west and south of the Chicago Loop, between one-fourth and one-half of all households have incomes below $25,000. One can travel south from downtown Chicago for 100 city

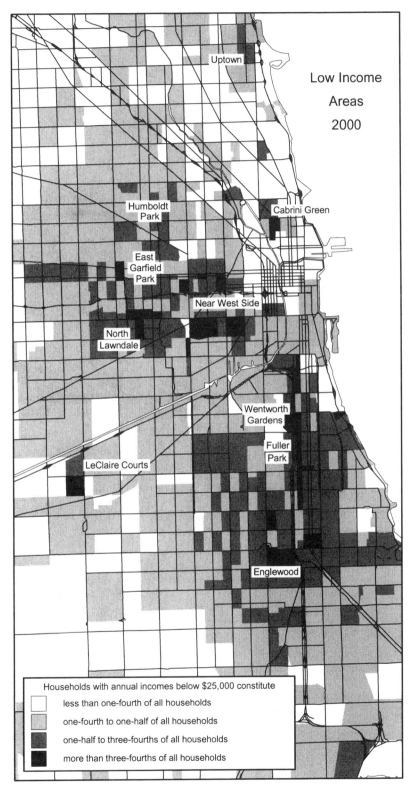

Low Income
Areas
2000

Uptown

Humboldt
Park

Cabrini Green

East
Garfield
Park

Near West Side

North
Lawndale

Wentworth
Gardens

Fuller
Park

LeClaire Courts

Englewood

Households with annual incomes below $25,000 constitute

- less than one-fourth of all households
- one-fourth to one-half of all households
- one-half to three-fourths of all households
- more than three-fourths of all households

Fig. 18.1

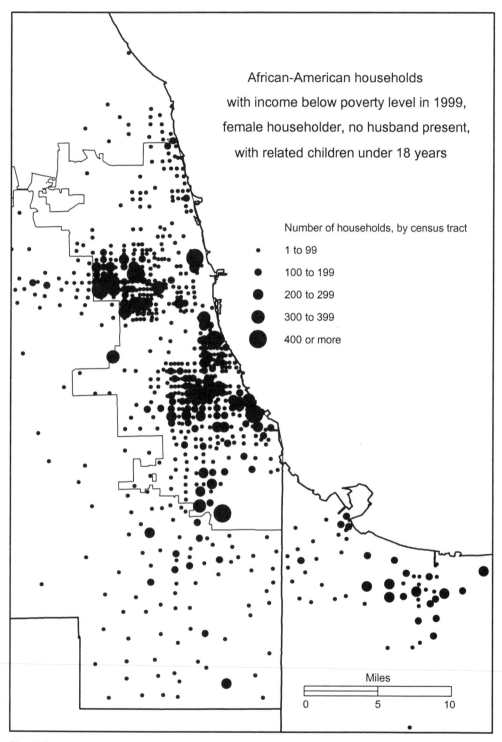

African-American households
with income below poverty level in 1999,
female householder, no husband present,
with related children under 18 years

Number of households, by census tract

· 1 to 99

● 100 to 199

● 200 to 299

● 300 to 399

● 400 or more

Miles

0   5   10

Fig. 18.2

blocks and not pass through more than a handful of census tracts that do not have either a majority or a substantial minority of families in the lowest-income categories. Much the same can be said of a traverse directly west from downtown. That this is still true, a century after the first "Bronzeville" neighborhood appeared on Chicago's South Side, has to be the most difficult to accept of any "fact" about the city's geography.

For a variety of reasons, mostly related to lack of available housing for poor families, concentrations of very low-income households are less obvious with distance away from the city's center. While poor families are less apt to constitute a substantial share of the population at greater distances from persistent poverty areas, poor families are nonetheless present in significant numbers beyond the Chicago city limits and into the suburbs.

Probably the most at-risk of all households are those described as African-American, with income below the poverty level, having a female household head, no husband present, and including one or more related children under eighteen years of age (Fig. 18.2). The densest concentration of such households coincides with the low-income areas shown in Fig. 18.1, but they are also scattered in all directions away from the hard-core poverty areas. Southern Cook County and adjacent portions of northwest Indiana have thousands of households in this category.

While most of these at-risk households number only in the tens or hundreds per census tract in outlying areas—rather than in the thousands, as is true of predominantly African-American portions of the central city—their appearance in the suburbs raises several issues. Moving to the suburbs obviously does not guarantee that a poor, dependent household will climb out of that status. Elimination of the worst inner-city public housing projects is now well underway in Chicago, but the map suggests that the conditions that gave rise to such projects have scarcely disappeared. Dispersing poor households over the metropolitan area is without a doubt better than concentrating them in the inner city; but moving to the suburbs neither eliminate such households nor does it solve the problems that they face.

HIGH-INCOME CONCENTRATIONS

High-income households are concentrated in several kinds of Chicago neighborhoods identified in previous chapters (Fig. 18.3). High incomes are associated with areas where employment in professional and managerial occupations is most common (Fig. 3.2), where white and Asian populations are increasing (Figs. 17.4 and 17.5), where new housing has been built (Fig. 17.6), and where people in the 20 to 29 age group live (Fig. 17.3). The high density of such households near the lakefront reflects both residential preference and the fact that those neighborhoods have a high overall population density.

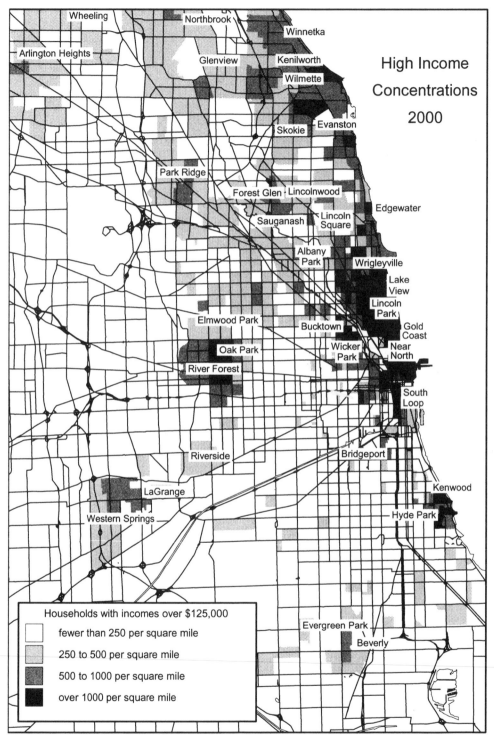

Fig. 18.3

Clusters of high-income households also are beginning to appear on Chicago's Northwest Side where neighborhood gentrification is underway. These are areas of single-family homes that are attractive to young professional couples whose first residence often was in the Lincoln Park, Lake View, or Wrigleyville areas. On the Northwest Side of the city, high-income household areas also are associated with reverse commuting (Fig. 16.2). Most such households have two members who hold full-time jobs.

A high-density pattern of affluent households continues north beyond the Chicago city limits and reemerges in Kenilworth, Wilmette, and parts of Evanston. Beyond these areas such households are less concentrated, not because of lower incomes but because of lower levels of population density. High-income households form a large percentage of all households in those suburbs (Fig. 18.4). On a community basis, the greatest concentration of high-income households is in southern Lake County and extreme northern Cook County. Oak Brook and Burr Ridge, in DuPage County, are the only western suburbs where the typical family has an income of more than $125,000 per year.

Away from the affluent North Shore suburbs, most communities where high incomes predominate are of relatively recent vintage. For a community to have more than half of its households in the $125,000-plus category means not only the presence of high-income families, but also the absence of poor ones. The smaller the population and the more irregularly configured the community's borders, the easier it is to maintain homogeneity. The cluster of high-income communities stretching from Barrington Hills to Lincolnshire constitutes a relative unbroken stretch of high-income residential developments. Most of the area was built up between 1970 and 1990 (Fig. 14.4). Residential enclaves there alternate with office parks and corporate headquarter campuses.

The sixteen high-income suburbs along the Cook-Lake county line had a combined population of roughly 75,000 in 2000. To the extent that Chicago has a zone of new, affluent communities, this is it, although neither the number of communities nor their total population is particularly large. Chicago lags far behind both New York City and Los Angeles in this respect. Both of those cities, while larger in total size, have extensive zones of high-income satellite communities that lie in various directions and sometimes at a substantial distance from the central city.

Nor is Chicago likely to acquire much more of this kind of development any time soon. The types of employment that give rise to high-income suburbs have become progressively confined to the northwest sector of Chicago's metropolitan area during the past several decades. It is difficult to imagine where else, around the several-hundred-mile perimeter of Chicago's outer suburban zone, such developments could

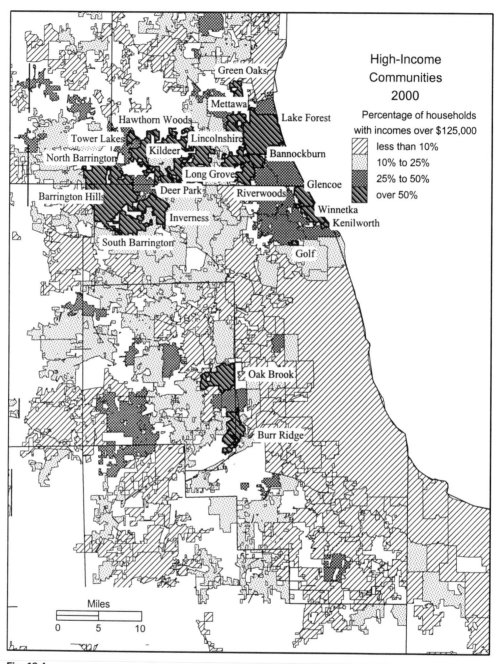

High-Income
Communities
2000

Percentage of households
with incomes over $125,000

- less than 10%
- 10% to 25%
- 25% to 50%
- over 50%

Green Oaks

Mettawa

Lake Forest

Hawthorn Woods

Tower Lakes
Lincolnshire

Kildeer

North Barrington

Bannockburn

Long Grove

Deer Park

Glencoe

Riverwoods

Barrington Hills

Winnetka
Inverness

Kenilworth

South Barrington

Golf

Oak Brook

Burr Ridge

Miles

0   5   10

Fig. 18.4

take place. As with many other trends in recent decades, the tendency in suburban residential patterns is toward heterogeneity and dispersion rather than homogeneity and concentration. This is especially true of most of the new suburban ethnic patterns (Figs. 15.11-15.14). High-income households are likely to continue to appear in many outlying areas, but the exclusivity of "high income only" seems to be inconsistent with other trends at present.

AFRICAN-AMERICAN SUBURBAN HOUSEHOLDS

Most African-American households in the Chicago area still are found in the central city, although there has been a sizable growth of black population in the suburbs during the past several decades. Even a few high-income suburban enclaves have appeared. Most African-Americans who live outside the city of Chicago, but within the metropolitan area, reside in one of the older satellites cities. Evanston, Waukegan, Aurora, Joliet, and Gary have had large black populations for many years (Fig. 18.5). Adjacent suburbs, such as Melrose Park and Maywood on the West Side and Harvey, Dolton, and South Holland on the South Side, have mixed populations that include a substantial African-American minority. Black populations in these smaller suburbs have grown substantially since the 1960s.

Nearly one-third of the 300 suburban communities in the greater Chicago region have no African-American-headed households at all. Most are communities with small populations lying some distance from the central city. Roughly five dozen communities have between one and ninety-nine black households. The most significant departure from past conditions is the nearly 100 villages and cities having 100 to 1,000 African-American households each. They are widely scattered over DuPage and northern Cook counties and, in fact, constitute the typical suburb in those areas.

The movement of more than just a few black families into these communities clearly did not trigger the "white flight" common in Chicago decades ago. These same communities absorbed hundreds of Asian and Hispanic families at the same time that African-Americans were moving in. There are no "black suburbs" in these areas, nor are there any "Indian suburbs," but thousands of black families, Asian families, and others now live in the same communities.

Viewed another way, a total of only 197 suburbs in the Chicago region have as many as twenty-five African-American-headed households each. Only five of them had black median household incomes above $125,000 in 1999. They are Frankfort Square, Ingalls Park, Plainfield, and Burr Ridge, Illinois, and Dyer, Indiana (Fig. 18.6). All are small communities and, except Dyer, are of fairly recent vintage. Another thirteen communities, ranging from Crystal Lake, Illinois, to Chesterton, Indiana, had median incomes of African-American households in the $100,000 to

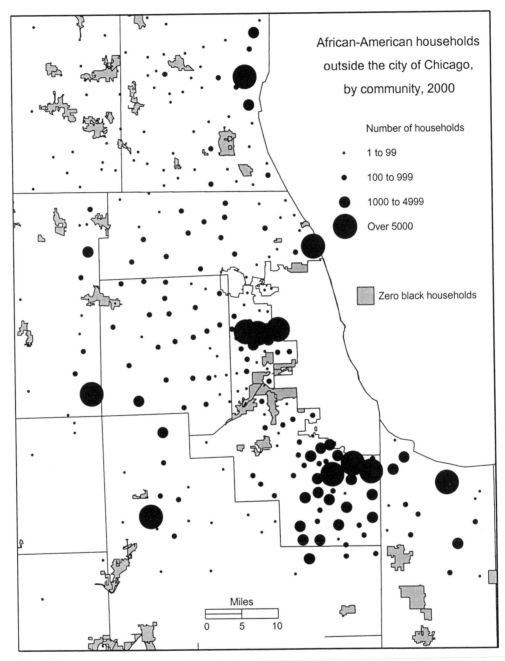

Fig. 18.5

$125,000 range on a base of at least twenty-five black households. All other communities with at least twenty-five black households are divided evenly at the $50,000 level: ninety are above and eighty-nine are below the value that is close to the $51,046 median for all households in the Chicago region.

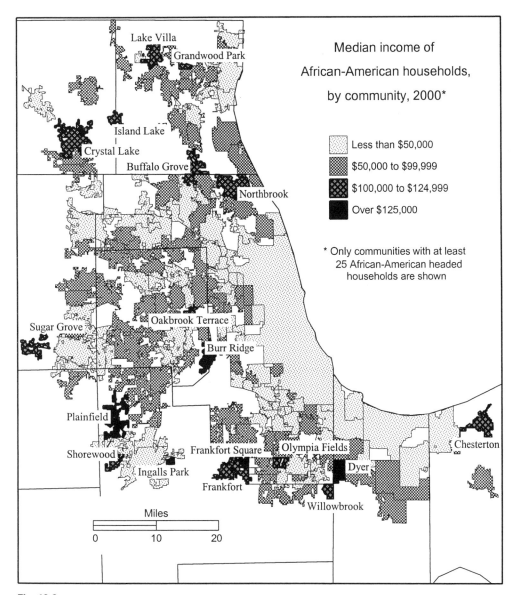

Fig. 18.6

These statistics point to the existence of a black middle class in the suburbs, if not to a sizable black wealthy class. The fact that middle- and higher-income African-American households are scattered over many communities and not confined to a few is perhaps the best evidence of an emerging break with past conditions. Although the convergence is thus far confined to upper-income levels, the residential patterns of African-American households finally are beginning to look more like those of other racial and ethnic groups.

Part VII

CONCLUSION

Chapter 19
THE SECTOR MODEL REVISITED

Nᴀᴛᴜʀᴇ ᴇɴᴅᴏᴡᴇᴅ Cʜɪᴄᴀɢᴏ with a site that presented few obstacles to urban expansion. The flat lake plain on which the city grew had poor natural drainage, but even this was a mixed blessing. Meandering streams were easy to redirect in order to speed the flow of surface waters. The natural flow of some streams changed direction during prehistoric times as water sought a more direct outlet through marshy lands to the lake. The configuration of the Chicago River at the time Euro-American settlement began consisted of two branches radiating symmetrically away from a short channel that carried their combined flow less than a mile into Lake Michigan. The city's first important economic activity was construction of the Illinois and Michigan Canal to breach the low drainage divide between the Illinois River basin and Lake Michigan.

The city's first plat was a rectangle of several dozen blocks in a grid of streets straddling the junction of the Chicago River's branches. From its beginning the city was thus divided into a North Side, South Side, and West Side by the arrangement of rivers. Economic activities clustered along the river banks. As the city grew, more commercial and industrial activities located along the rivers. A specialization in land use emerged by the 1860s with meat packing, lumber wholesaling, and grain shipping—three of Chicago's important early industries—evolving in specialized clusters in various directions (Fig. 8.2).

Railroads were added one by one from the late 1840s to the end of the nineteenth century. Each line entered the city from a slightly different direction, which provided a new basis for sectoral alignments. Industrial land uses migrated outward along the railroad lines, and residential developments filled in the wedges between them. Land-use intensity decreased with distance away from the center, as in all cities, and varied substantially in its nature and intensity between the radiating sectors.

Deliberate attempts at city planning, including the designation of parks, boulevards, and other thoroughfares, occurred after the Chicago fire of 1871 forced a rebuilding. Parks and other amenities were distributed between the North, West, and South sides of the city (Fig. 9.1). Portions of the lakefront were filled in and the strip of lakefront park land began to emerge. Growth spurts alternated between various sides of the city. The 1890s was a period of substantial expansion southward, especially in connection with the World's Fair of 1893. Population was distributed quite evenly over the city's residential areas in 1900, although areas of only sparse settlement remained near the northwest and southeast city limits (Fig. 9.4).

In 1909, Daniel Burnham and Edward Bennett published their *Plan of Chicago*. Popularly known as the Burnham plan, it called for dramatic changes, especially in the downtown area (Fig. 10.1). The boldest proposal was for a new Civic Center, to be built at the intersection of Congress and Halsted streets, which was then near the center of population in the city. Six boulevards radiating symmetrically from the Civic Center would have made it the focal point of Chicago. The existing grid of city blocks had to be retained as a practical matter, but subsequent thoroughfares were to be mostly radial.

The Burnham plan revived classical ideas of urban form, including the incorporation of space as a formative element of the urban landscape. Symmetry, axiality, and balance were primary because they allowed a central focus. Burnham thought the city should be a "complete organism," a single unit, and thus it had to have a central core. The Burnham plan also included proposals for an outlying system of radial highways, a fringe of industrial districts, and strips of park land. Symmetries, sectors, and concentric rings abounded in his "City Beautiful/Garden City" scheme.

The Burnham plan did not demand that Chicago be radically restructured in general, nor would that have been necessary. The city had been growing in a geographically balanced fashion for more than half a century at the time Burnham made his proposals. The 1890s had seen perhaps the greatest step of all toward centralization and balance when the radiating routes of rapid transit trains were linked to a loop of track surrounding the central business district, permitting continuous operation of trains without turning. The Loop would define downtown Chicago from that time forward.

Centrally focused, balanced, radial, and symmetric structures share in common a geometric principle known as self-similarity. A circle is the most self-similar of all structures in the plane because it can be rotated infinitely many ways and still remain a circle. A square has more self-similarities than a rectangle, but a hexagon improves upon the square. Nature favors self-similar structures because they are efficient. Soap bubbles, ripples on a pond, veins in a leaf, growth rings in a tree, snowflakes, and salt crystals are some examples of natural self-similarity.

These ideas also can be applied to the structure of a city. Broadly speaking, self-similar structures are also democratic structures. Placing resources as close to as many people as possible is one definition of a democratic arrangement. Although access is not guaranteed, no one is inherently disadvantaged in a symmetric arrangement. The center is equidistant from every point on the radius of a circle. If the circle is eccentric, then access to the center is easier from some directions than it is from others. All other things equal, balanced designs offer the greatest degree of access for the largest numbers of people. Asymmetries offer advantage to some and disadvantage to others.

Chicago's development was geometrically balanced at the beginning of the twentieth century, even though specialized-function districts had emerged in numerous locations around the city. Shortly after the Burnham plan was unveiled, a major episode of unbalancing began that would overshadow the city's rational growth for the rest of the twentieth century. African-Americans appeared in significant numbers among new arrivals to the city beginning in the 1910s. Many Europeans had entered the city by establishing a first residence in one of the central immigrant ghettoes. African-Americans did the same, although the areas into which they were channeled were separate from those where Europeans lived. From its early beginnings south of downtown Chicago, the black neighborhoods spread straight south and, to a lesser extent, west from the city center.

Although these were sectoral expansions that, in some respects, were not greatly different from those made by, say, Polish immigrants in Chicago, African-Americans encountered much different circumstances. Adjacency to a black neighborhood caused whites to move elsewhere, which meant that African-Americans expanded into adjacent areas but rarely anywhere else. Despite public housing programs and good-faith efforts made by many, this established sequence continued unabated until much of the city's South Side and parts of its West Side had a residential composition that was almost entirely African-American.

As more whites moved to the suburbs, they tended to move in any direction that was not in the probable expansion path of the African-American population. Chicago's "bedroom communities" in the 1950s were scattered around the urban fringe in a fairly balanced fashion, but over time a drift toward the west and northwest became more evident.

Racial avoidance had something to do with the skewed pattern, but the unbalancing of the metropolis was equally the product of economic change. Chicago's manufacturing industries began to gravitate toward the northwest part of the city and adjacent suburbs in the 1950s. Construction of O'Hare Airport and completion of the Edens and Kennedy expressways aided the northwestward tilt. The shift to the northwest was especially evident in the newer industries, including electronics and high-value-added light manufacturing. Chicago's heavy industries, concentrated in the Calumet district on the Southeast Side, began to experience difficulties in the 1960s as steel production declined. Industrial growth made a net shift northwestward as a result.

"Economic restructuring," a phrase commonly used in the 1980s, involved the decline of many traditional manufacturing industries that had been a mainstay of Chicago's economy for years. Because the South Side of the city had been home to a large share of Chicago's manufacturing capacity in the past, the South Side disproportionately absorbed the effects of restructuring. Downsizing and corporate reorganization led to the closure of some manufacturing plants. Restructuring led to reduced employment in a wide range of industries that employed skilled labor. Eventually there was a decrease in demand for goods and services that had supplied those industries, which only produced another round of economic adjustments.

THE REVISED SECTOR MODEL

By the early twenty-first century, Chicago had evolved a new sectoral pattern (Fig. 19.1). The model is designed to be illustrative (rather than definitive) and is offered for the purpose of summarizing a variety of changes that have taken place. The categories are: "expansion growth," which refers both to population and economic growth in areas of expanding commercial and residential development; "maintenance growth," areas that are growing at an average rate for the metropolitan area; and "slow growth" and "decline," the areas most disadvantaged by the general northwestward drift in the metropolitan economy.

Of the several idealized types of city structure reviewed in Chapter Two, the sector model continues to best describe Chicago and its region. In the revised model, sectors are defined with respect to growth or change rather than in terms of land use, but the sector pattern remains evident (Fig. 19.1). Sectoral growth need not lead to asymmetry if all sectors are expanding equally, but such clearly has not been the case in Chicago in recent decades.

Even the effects of racial avoidance and economic restructuring need not have led to an unbalanced sectoral pattern had the metropolitan area as a whole been growing at a brisk pace. If all of Chicago's various sectoral wedges had continued to grow in proportion, as they did up through the beginning of the twentieth century, the metropolitan area would now have a total population and economic base equal to or greater than that of Los Angeles, which became the nation's second largest metropolis during the era of Chicago's relative decline. Put in different terms, a narrowing wedge of growth probably is more the result of external, rather than internal, influences.

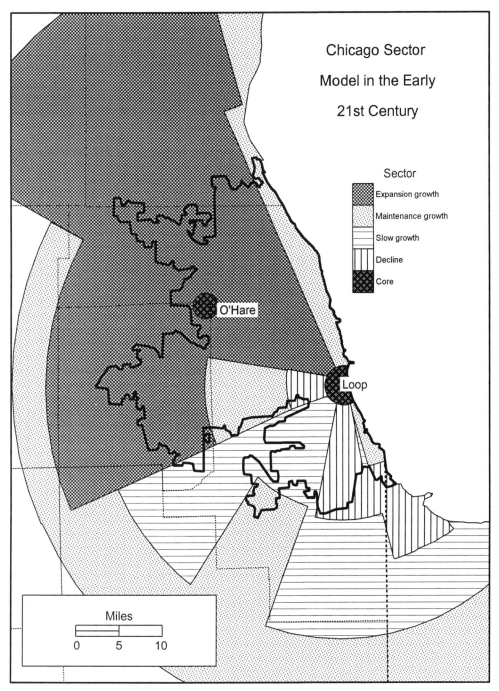

Fig. 19.1

THE NORTHWEST SECTOR

Expansion growth characterizes the northwest sector of the metropolitan area, which is the segment that continues to grow in a manner similar to the late nineteenth and early twentieth centuries when Chicago pioneered new forms of urban economic activity. The term "expansion growth" is deliberately broad, but it is supported by six specific lines of evidence regarding changes during the past half century.

DIRECTION OF SUBURBANIZATION. The growth of suburbs around Chicago shifted from a fairly even directional dispersion immediately after World War II to an increasingly sectoral focus on the northwest quadrant by the 1970s (Table 14.1). Early suburbs continued an older growth pattern in which newly built-up areas were nearly always contiguous to existing communities (Fig. 14.2). Suburbanization pushed the outer perimeter of new developments at a more rapid pace than population growth demanded, however, so that by the 1970s the pattern of new communities left many gaps (Figs. 14.3 and 14.4). The overall rate of growth in the metropolitan area was slow enough that new developments could be concentrated in just a portion of the suburban fringe. Both a sectoral growth bias and scattered development ("sprawl") were hallmarks of a metropolis expanding in area faster than it expanded in population.

NEW ECONOMIC ACTIVITIES. The distribution of firms classified in the professional, scientific, and technical category is strongly oriented to the northwest quadrant of the metropolitan area (Fig. 16.3). Chicago is less well known for these activities than are, say, Boston and San Francisco, but it has attracted a share of the growth because of specialization in products such as cellular telephones. Because this category includes a very large proportion of new companies, especially those commonly labeled as "hi-tech," it is understandable that they have located primarily on the suburban fringe and in those areas most identified with new growth. Such companies are often regarded as preferring suburban locations, and many do, but the northwest quadrant of the central city also is home to many such employers. The sectoral wedge begins in the Loop and fans out to the northwest. DuPage County is a separate aggregation. There is a glaring absence of such firms south of downtown Chicago (Fig. 16.3).

REVERSE COMMUTING. Suburbanization of workplaces was both a result and a cause of unbalancing in the metropolitan area. Jobs located in the central city are within fairly easy access to anyone in the metropolitan area, partly because public transportation systems are based almost entirely on commuting from suburban residence to central-city workplace. Placing those same jobs at scattered locations on the suburban fringe pushes them out of reach for many people. Reverse commuting, including a shift from public transportation to private automobile as the mode of choice, is largely a response to this circumstance.

Chicagoans who commute to jobs outside the city are strongly clustered on the Northwest Side, in those areas most associated with employment in professional, scientific, and technical firms (Fig. 16.2). High-income professionals along the lakefront also are reverse commuters, although they remain more likely to be employed in downtown Chicago than in the suburbs.

GENTRIFICATION. Renewed investment in older, inner-city neighborhoods often accompanies an increase in white (and, more recently, Asian) population. Spectacularly large percentage increases in white population can appear simply for statistical reasons in neighborhoods that were formerly all African-American in composition. A substantial increase in white and Asian population density is associated with neighborhoods where the turnover has produced significant numbers of new residents (Figs. 17.4 and 17.5). Gentrification typically involves young, higher-income households, and it is most in evidence along public transportation corridors. It has recently begun to expand away from those areas, especially along the Kennedy Expressway on the Northwest Side.

Thousands of city blocks in the city of Chicago are potential candidates for gentrification, of course, and new nodes could appear anywhere in the city in the years to come. Gentrification's strong association with double-income households, professional-level employment, and college-educated householders suggests that its greatest future growth is likely to follow from these associations and be most in evidence in a spread to the north and west.

NEW ETHNIC PATTERNS. Historically, Chicago's ethnic groups were identified with dozens of separate neighborhoods scattered about the city where people of common national origins settled. This pattern continues today to some extent, but new immigrants are more likely to reside in the suburbs than in the central city. Accompanying the shift to suburban residence is a marked decrease in the tendency for persons of similar national origins to cluster together. The trend is most evident among Asian immigrant groups (Figs. 15.11 and 15.14). Their employment patterns are strongly associated with professional, scientific, and technical firms and are centered in suburban areas west and north of the city where those firms have located.

Mexico and Poland are the two largest country sources of immigrants to the Chicago area today (Figs. 15.6 and 15.9; Table 15.1). Both nationalities have moved into the Northwest Side of Chicago and to its northwest suburbs in substantial numbers. The sheer volume of migration for these two groups is sufficient to provide a scatter over many portions of the metropolitan area, but they, too, appear to follow the trend evident for other groups.

INCOME. All of Chicago's high-income suburbs are located west, northwest, or directly north of the city (Fig. 18.4). While the older, high-status North Shore suburbs remain part of this group, the largest concentration of high-income communities lies in a thirty-mile strip from Lake Forest to the southeastern corner of McHenry County.

Many lakefront neighborhoods on Chicago's North Side are also in the high-income category and show up disproportionately as areas of white-collar employment (Fig. 3.2). This is an almost exclusively residential zone whose residents are either reverse commuters (Fig. 16.2) or users of public transportation (Figs. 16.7 and 16.8), which means daily work trips to downtown Chicago. Because it is a highly valued residential area, it is unlikely to become an area of new workplace growth in the future.

African-Americans still are substantially underrepresented in the high-income categories, although middle- and upper-income black families have a residential pattern that now more closely resembles that of the population in general (Figs. 18.5 and 18.6). Among upper-income groups, only African-Americans are just as likely to reside in the South Side suburbs as they are in those to the north and west. Their concentration of residence on the South Side means that African-Americans have farther to commute to new jobs available in the expanding northwest sector. Inequalities in one era thus are perpetuated in the next.

OVERALL TREND

These six indicators confirm a revised sectoral-model interpretation of Chicago that emphasizes a net expansion in the metropolis toward the north and west and a corresponding slackening of growth on the south and southeast (Fig. 19.1). Thus far, at least, this tendency has remained anchored in downtown Chicago. The Loop continues to be the economic heart of the city and the largest place of employment for residents living directly to its north or south. The O'Hare Airport vicinity constitutes a secondary core that depends heavily on downtown for its own support. The northwestward tilt emphasized in the revised model is thus more of a stretching than a net relocation.

Whether a continuation of these trends will lead to a "stretched" Chicago, aligned more noticeably along a northwest axis than around a single downtown focus, must be an issue that will confront regional planners and public officials in the years ahead. Transportation systems have granted slight recognition to the trend thus far, although people and workplaces have been following it for decades. Chicago's future geography could look very different from that of the past. Whether that future is guided by a design that serves the metropolis as well as Chicago's early structure did is an issue for all to consider.

Chicago Portfolio

WHERE GEOGRAPHY AND PHOTOGRAPHY MEET

EDITED AND CURATED BY GEORGE F. THOMPSON

A signed limited edition of 250 copies of *Chicago Portfolio: Where Geography and Photography Meet* (ISBN: 1-930066-45-7) was published by the Center for American Places, of Santa Fe, New Mexico, and Staunton, Virginia, to commemorate the annual meetings of the Association of American Geographers and the Society for Photographic Education, held respectively in Chicago in March 2006. Once again, Chicago brings together the worlds of art and scholarship.

Published 2006. First Limited Edition.
Printed in China. The text was set in Granjon and Univers. The paper is Chinese Goldeast, 157 gsm weight.
David Skolkin, of Santa Fe, book designer and art director.
Dave Keck, of Global Ink, Inc., production coordinator.

The Center for American Places, Inc.
P.O. Box 23225
Santa Fe, New Mexico 87502, U.S.A.
www.americanplaces.org

INTRODUCTION
by George F. Thompson

Chicago has long been home to some of the nation's most esteemed and accomplished photographer-artists. Among them are Harold Allen, Harry Callahan, Barbara Crane, Yasuhiro Ishimoto, Joe Jachna, Ray K. Metzker, Arthur Siegel, Art Sinsabaugh, Aaron Siskind, Joseph Sterling, and the recently discovered Gary Stochl. Nonresident artists, such as Walker Evans and John Szarkowski, were also attracted to the city, making important and unforgettable photographs of Chicago, its people, and environs. Other photographers such as Richard Nikel have tirelessly recorded, and helped to preserve, the city's rich architectural legacy, just as numerous contemporary photographers (including the ten featured in this portfolio) continue to interpret Chicago's changing sense of place through their work. Significant collections of fine-art and historic images reside at the Art Institute of Chicago, Chicago Historical Society, Chicago Landmarks Commission, Chicago Park District, LaSalle Bank, and the Museum of Contemporary Photography at Columbia College Chicago, among others. Each of these institutions shares a commitment to photography—as an art form, to be sure, but also as a visionary indicator of the documentation and various meanings of Chicago's people, places, buildings, and outdoor spaces. It is no wonder that Chicago provides such an ideal intersection of geography and photography.

The photographs in *Chicago Portfolio* were selected either from books that appear in two of the Center's many book series, *Center Books on Chicago and Environs* (since 2001) and *Creating the North American Landscape* (1989–2006), or from projects that are in accord with the objective of those series: to enhance the public's understanding of, appreciation for, and affection for the natural and built environment through art, literature, and scholarship.

The photographs in *Chicago Portfolio* represent a continuation of Chicago's photographic tradition mentioned above, in which Chicago is a place that not only nurtures and inspires creative, descriptive, and insightful photographic art, but also makes possible the necessary connection between geography and photography. That connection has long been a hallmark for scholars, artists, and writers who are compelled to study and interpret places—urban, rural, suburban, and wild—wherever they may be.

ACKNOWLEDGMENTS AND CREDITS

Special thanks to John C. Hudson, author of *Chicago: A Geography of the City and Its Region* (2006), for believing in the value and appropriateness of including *Chicago Portfolio* in his book; to Julia Sniderman Bachrach, author of *The City in a Garden: A Photographic History of Chicago's Parks* (2001), for making possible the reproduction of the photographs and captions from her book; to Walker Blackwell, Vice President of Black Point Editions in Chicago, for his pre-press work on the photographs by Terry Evans; to the Chicago Park District, for permission to use historic photographs from its Special Collection; to Greg Foster-Rice, Assistant Professor of Art History in the Photography Department at Columbia College Chicago, for his editorial counsel regarding the history of photography in Chicago; to Benjamin Gest, Director of Digital Scans at Columbia College Chicago, for his pre-press work on the photographs by Paul D'Amato, Scott Fortino, Gina J. Grillo, Melissa Ann Pinney, Brad Temkin, Bob Thall, and Jay Wolke; to Paul Lane, of Evanston, Illinois, for his pre-press work on the historic images from the Chicago Park District's Special Collection; to Amber K. Lautigar, Ashleigh A. Frank, and Brian M. Venne, at the Center for American Places, for preparing the manuscript for publication; to Randall B. Jones, who for years helped the Center for American Places to advance the concept of "photography and geography" in the Center's books; and to all of the artists featured, for making *Chicago Portfolio* possible through their generous spirit and commitment to understanding *place*.

THE ARTISTS

JUDITH BROMLEY	PLATES 31–32
PAUL D'AMATO	PLATES 49–56
TERRY EVANS	PLATES 79–84
SCOTT FORTINO	PLATES 43–48
GINA J. GRILLO	PLATES 15–21
JIM ISKA	PLATES 34–37, 39, AND 41–42
MELISSA ANN PINNEY	PLATES 22–28
BRAD TEMKIN	PLATES 64–70
BOB THALL	PLATES 1–14 AND 71–78
JAY WOLKE	PLATES 57–63

Historic photographs from the Chicago Park District's Special Collection
appear as Plates 29, 30, 33, 38, and 40.

Gallery I

DOWNTOWN CHICAGO, 1978–2004

PHOTOGRAPHS BY BOB THALL

BOB THALL (b. 1948, Chicago) is Professor and Chairman of the Photography Department at Columbia College Chicago. He is the author of *The Perfect City* (1994), which was accompanied by a one-person show at the Art Institute of Chicago, *The New American Village* (1999), which was accompanied by a one-person show at the Museum of Contemporary Photography in Chicago, *City Spaces: Photographs of Chicago Alleys* (2002), and *At City's Edge: Photographs of the Chicago Lakefront* (2005). He has received a John Simon Guggenheim Memorial Foundation Fellowship for Photography, and his photographs are in the permanent collections of the Art Institute of Chicago, Canadian Centre for Architecture, Getty Center for the History of Art and Humanities, Library of Congress, Museum of Fine Arts in Houston, and Museum of Modern Art, among others. Bob Thall's photographs in Gallery I of *Chicago Portfolio* come from *The Perfect City* (Plates 1–8 and 14), *City Spaces* (Plate 9), and *At City's Edge* (Plates 10–13).

PLATE 1: View west, Lake Shore Drive at Lake Street, 1978.

PLATE 2: View east, the Chicago River at the split of the North and South branches, 1982.

PLATE 3: View south, the Chicago River from beneath the Merchandise Mart, 1984.

PLATE 4: View east, Wacker Drive from Wabash Avenue, 1978.

PLATE 5: View north, Franklin Street at Lake Street, 1979.

PLATE 6: View northwest, Lake Street east of Michigan Avenue, through the Illinois Center, under construction, 1980.

PLATE 7: View south, Columbus Drive at East Wacker Drive, 1979.

PLATE 8: View north, Franklin Street at Lake Street, 1991.

PLATE 9: The alley between Wells Street and Franklin Street, in the vicinity of Washington Street, 1998.

PLATE 10: View west, toward downtown, from the Adler Planetarium, 2002.

PLATE 11: View north, toward downtown, from the Adler Planetarium, 2002.

PLATE 12: View south, Lake Michigan near Addison Street, 2004.

PLATE 13: View east, Lake Michigan near Erie Street, 2004.

PLATE 14: View east, from the roof of the IBM Building, 1989.

Gallery II

COMMUNITY AND NEIGHBORHOOD LIFE, 1996–2004

PHOTOGRAPHS BY GINA J. GRILLO AND MELISSA ANN PINNEY

GINA J. GRILLO (b. 1960, Lake Forest, Illinois) has taught photography as an adjunct faculty member in the Photography Department at Columbia College Chicago since 1997. She had a one-person show at the Ellis Island Immigration History Museum and Statue of Liberty Monument in New York City to coincide with the publication of *Between Cultures: Children of Immigrants in America* (2004). Her photographs are in the collections of Albert and Tipper Gore, the Fannie Mae Foundation, and the University of Chicago, among others. Gina Grillo's photographs in *Chicago Portfolio* (Plates 15–21) come from *Between Cultures*.

MELISSA ANN PINNEY (b. 1953, St. Louis, Missouri) has taught photography at Columbia College Chicago since 1984. She has received fellowships in photography from the John Simon Guggenheim Memorial Foundation, National Endowment for the Arts, and Illinois Arts Council, and her photographs are in the permanent collections of the Art Institute of Chicago, Center for Creative Photography, Metropolitan Museum of Art, Museum of Modern Art, San Francisco Museum of Modern Art, and Whitney Museum of American Art. She is the author of *Regarding Emma: Photographs of American Women and Girls* (2003). Melissa Pinney's photographs in *Chicago Portfolio* (Plates 22–28) come from her ongoing project, *Feminine Identity*, begun in 1987, including one photograph (Plate 23) from *Regarding Emma*.

PLATE 15: Chicago Trinity Irish Dancers, along Lake Michigan, 1996.

PLATE 16: Watching a parade on Good Friday, Pilsen, 1997.

PLATE 17: Shoppers at Devon and Western, West Rogers Park, 1997.

PLATE 18: St. Lucia celebration, Swedish American Cultural Center, Andersonville, 1997.

PLATE 19: Polish procession, Ukrainian Village, 2002.

PLATE 20: Pledge of Allegiance, Walt Disney Magnet School, North Side, 1997.

PLATE 21: Students from Africa and Asia, St. Thomas of Canterbury School, Uptown, 1999.

PLATE 22: Taiwanese Independence Day, Chinatown, 2004.

PLATE 23: Ice cream social, Evanston, 2001

PLATE 24: Baseball game, Leahy Park, Evanston, 2003.

PLATE 25: Croquet game at a retirement home, Evanston, 2003.

PLATE 26: Halloween, Curry Park, Evanston, 2004.

PLATE 27: Curry Park, Evanston, 2004.

PLATE 28: Millennium Park, 2004.

Gallery III

PUBLIC PARKS, 1890–2000

PHOTOGRAPHS BY JUDITH BROMLEY AND JIM ISKA
AND HISTORIC PHOTOGRAPHS FROM THE CHICAGO PARK DISTRICT'S
SPECIAL COLLECTION

JUDITH BROMLEY (b. 1944, Morristown, New Jersey) is a photographer/artist who lived and worked in Chicago from 1983 to 2004. Her photographs are in numerous private collections, have appeared in such magazines as *Architectural Digest*, *Architecture*, *Atlantic Monthly*, *Garden Design*, *Historic Preservation*, *House Beautiful*, and *Progressive Architecture*, and were featured in the *AIA Guide to Chicago Architecture* (1993) and *The City in a Garden: A Photographic History of Chicago's Parks* (2001). As a photographer, she is co-author, with Kathryn Smith, of *Frank Lloyd Wright's Taliesin and Taliesin West* (1977). Judith Bromley's photographs in *Chicago Portfolio* (Plates 31–32) come from *The City in a Garden*.

JIM ISKA (b. 1958, Chicago) joined the Art Institute of Chicago in 1982, where he is a departmental specialist in the Department of Photography. His photographs are in the permanent collections of the Canadian Centre for Architecture, Chicago Historical Society, Museum of Contemporary Photography, and Museum of Fine Arts in Houston, among others. As a photographer, he is co-author, with Francis Morronen, of *An Architectural Guidebook to Brooklyn* (2001), *An Architectural Guidebook to New York City* (1998), and *An Architectural Guidebook to Philadelphia*. He was also a contributing photographer to *Changing Chicago: A Photodocumentary* (1989) and *The City in a Garden: A Photographic History of Chicago's Parks* (2001). Jim Iska's photographs in *Chicago Portfolio* (Plates 34–37, 39, and 41–42) come from *The City in a Garden*.

The photographs from the Chicago Park District's Special Collection in *Chicago Portfolio* (Plates 29, 30, 33, 38, and 40) come from *The City in a Garden*.

PLATE 29: Meant to emulate a natural "prairie river," Jens Jensen edged his meandering waterway in Humboldt Park with aquatic plants such as arrow root, cattails, pickerelweed, and water lilies. The park was named in honor of Baron Friedrich Heinrich Alexander Von Humboldt, the famous German scientist, naturalist, and explorer. (1941.)

PLATE 30: Washington Park's landscape, designed by Frederick Law Olmsted, featured elegant, naturalistic spaces for relaxation and contemplation. Originally called South Park, the Midway Plaisance connected its two divisions: the western division was renamed Washington Park in 1881 in honor of George Washington, the first president of the United States; the eastern division became known as Jackson Park in 1880 in honor of Andrew Jackson, the nation's seventh president. (ca. 1890.)

PLATE 31: The recent restoration of Lincoln Park's North Pond includes prairie plantings along the water's edge. Originally known in 1860 as Lake Park, the site (a former lakeside public cemetery) was renamed Lincoln Park shortly after the assassination of Abraham Lincoln, the nation's sixteenth president. (2000.)

PLATE 32: During the late 1990s, the Museum Campus Project resulted in major improvements to both Grant and Burnham parks. This area, once a vast surface parking lot just north of the Field Museum of Natural History, now offers beautiful terraced gardens and dramatic views of Lake Michigan and the Chicago skyline. Burnham Park honors Daniel H. Burnam, Chief of Construction for the World's Fair of 1893 and author of the famous *1909 Plan of Chicago*; Grant Park (known as Lake Park from 1847-1901) honors Ulysses S. Grant, the eighteenth president of the United States. (2000.)

PLATE 33: A large majority of Lincoln Park's current 1,208 acres were created through landfill operations conducted between the 1880s and 1950s. This aerial view looks southward from Foster Avenue. The largest park in Chicago, this linear landscape includes six bathing beaches, three boating harbors, a golf course and driving range, dozens of monuments and sculptures, a conservatory and gardens, the Chicago Historical Society, and numerous other cultural and recreational features. (1938.)

PLATE 34: When it opened to the public in 1908, Jens Jensen's Garfield Park Conservatory was referred to as "landscape art under glass." The premier space, the Fern Room, was inspired by the prehistoric landscape of Illinois, featuring a wonderful composition of water, rock, and plants. Originally known as Central Park, Garfield Park was conceived in 1869 as the center-piece of the West Park System and renamed in 1881 after the assassination of President James A. Garfield. (2000.)

PLATE 35: Shortly after the Illinois State Legislature formed the West Park System in 1869, the southernmost part was named in honor of Senator Stephen A. Douglas of Illinois, famed orator who brought the Illinois Central Railroad to Chicago. Douglas Park's original great meadow is now used as soccer fields. (1995.)

PLATE 36: Sherman Park, designed by the Olmsted Brothers and D. H. Burnham & Company in 1904, has four historic bridges that cross its lagoon to ball fields on the island. The park was named for John B. Sherman, founder of the nearby Union Stock Yards, who served as a South Park commissioner for twenty-five years. (1995.)

PLATE 37: Marquette Park pays tribute to Father Jacques Marquette, the famous French Jesuit missionary and explorer. In 1904, the Olmsted Brothers developed an impressive plan for the park. Today, its lagoons are extremely popular for fishing and are stocked with blue gill in the spring and channel catfish in the summer. (2000.)

PLATE 38: Union Park was created in 1853 and named for the Federal Union. In 1888, Jens Jensen planted his experimental *American Garden* there, marking the beginnings of his venerable naturalistic style. In 1900, African Americans began to move into the neighborhood. While many other Chicago parks were inaccessible to black residents, Union Park became racially integrated. In 1920, forty percent of Union Park's patrons were black and sixty percent were white, quite unusual for the time. (ca. 1920.)

PLATE 39: A collaborative effort between the Chicago Park District, Chinatown community, and Chinese-American landscape architect Ernest Wong, the new Ping Tom Memorial Park includes an entry plaza, children's playground, gardens that feature Chinese landscape design, and a riverfront pavilion at the edge of the South Branch of the Chicago River. The twelve-acre park honors Ping Tom, a life-long Chinatown resident who was an advisor to U.S. senators, Illinois governors, and Chicago mayors. The site was originally a Chicago and Western Indiana Railroad yard. (2000.)

PLATE 40: During the early twentieth century, the West Park Commission managed its landscapes with great sensitivity to nature. Laborers spent several winter days digging and transplanting this mature tree in Humboldt Park to its new site near the park's boat house. (1907.)

PLATE 41: Originally displayed in plaster on a Garfield Park garden in 1909, sculptor Edward Kemeys's bison was recast in bronze and installed in Humboldt Park's rose garden two years later. Kemeys, who specialized in sculptures of animals, was also responsible for the famous pair of lions that flank the Art Institute of Chicago. (2000.)

PLATE 42: The gift of Kate Buckingham in honor of her brother Clarence, the Buckingham Memorial Fountain in Grant Park was the world's largest decorative fountain when it was dedicated in 1927. It remains a prominent and popular downtown feature. (2000.)

Gallery IV

PUBLIC HOUSING, 1998–2001

PHOTOGRAPHS BY SCOTT FORTINO

THE WEST SIDE, 2003–2005

PHOTOGRAPHS BY PAUL D'AMATO

SCOTT FORTINO (b. 1952, Chicago) is a photographer/artist who has worked as a police officer with the Chicago Police Department since 1980. His photographs are in the permanent collections of the LaSalle Bank of Chicago and Museum of Contemporary Photography, where he had a one-person exhibition in 2002. He is the author of *Institutional: Photographs of Jails, Schools, and Other Chicago Buildings* (2005). Scott Fortino's photographs in *Chicago Portfolio* (Plates 43–48) come from his photographic project, *Landscape with Intent: The Demolition of Chicago's Public Housing*.

PAUL D'AMATO (b. 1956, Boston, Massachusetts) is a professor of photography at Columbia College Chicago. He has received a fellowship in photography from the John Simon Guggenheim Memorial Foundation, and he is the author of *Barrio: Photographs from Chicago's Pilsen and Little Village* (2006). His photographs are in the permanent collections of the Museum of Modern Art, Museum of Contemporary Photography, Metropolitan Museum of Art, and Fogg Museum at Harvard University. Paul D'Amato's photographs in *Chicago Portfolio* (Plates 49–56) come from his ongoing photographic work on Lake Street and Chicago's West Side.

PLATE 43: View of downtown Chicago from a Cabrini-Green Extension building, 1121 North Larrabee Street, 2001.

PLATE 44: The Stateway Gardens housing project, awaiting demolition, along South State Street near 37th Street, 2000.

PLATE 45: View of St. Ignatius Church and the demolition of ABLA Homes,
along Roosevelt Road near Blue Island, 1999.

PLATE 46: Looking south at Cabrini-Green Extension buildings on North Larrabee Street, at the intersection of Clybourn Street, 2001.

PLATE 47: A Robert Taylor Home building, ready for demolition, near 46th Street and South Federal Street, 1999.

PLATE 48: View of downtown Chicago from 41st Street and LaSalle Avenue, 1998. On the left is the Dan Ryan Expressway; in the middle are the METRA and Amtrak rail lines; and to the right of the tracks are high-rises from Stateway Gardens in the center background and a Robert Taylor Home building being demolished on the far right.

PLATE 49: View west from the 8th floor of a Henry Horner Homes building, 2005.

PLATE 50: Pastor of the Apostle Point Non-Denominational Church, 319 North Pulaski Road, 2003.

PLATE 51: PT at Cabrini-Green, 2005.

PLATE 52: Girl with laundry on Lake Street, 2004.

PLATE 53: Boy at Red Hots on Lake Street, 2003.

PLATE 54: Sunday at the Holy Temple of God Community Church, 4504 West Harrison Street, 2005.

PLATE 55: Boy at the Union Park pool, 2005.

PLATE 56: Too Tall, in front of the remains of the Henry Horner Homes housing project, 2005.

Gallery V

THE EXPRESSWAY AND PRIVATE PLACES, 1981–2004

PHOTOGRAPHS BY JAY WOLKE AND BRAD TEMKIN

JAY WOLKE (b. 1954, Chicago) is a professor in the Art and Design Department at Columbia College Chicago. He is the author of *Along the Divide: Photographs of the Dan Ryan Expressway* (2004) and *All Around the House: Photographs of American-Jewish Communal Life* (1998), which was accompanied by a one-person show at the Art Institute of Chicago. His photographs are in the permanent collections of the St. Louis Museum of Art, Museum of Modern Art, Museum of Contemporary Photography, Minneapolis Museum of Art, Brooklyn Museum, and Art Institute of Chicago, among others. Jay Wolke's photographs in *Chicago Portfolio* (Plates 57–63) come from *Along the Divide*.

BRAD TEMKIN (b. 1956, Chicago) has taught photography at Columbia College Chicago since 1984. He is the author of *Private Places: Photographs of Chicago Gardens* (2005). His photographs have been exhibited widely throughout the United States, and they are in the permanent collections of the Art Institute of Chicago, Krannart Art Museum, Museum of Contemporary Photography, and Museum of Fine Arts in Houston, among others. Brad Temkin's photographs in *Chicago Portfolio* (Plates 64–70) come from *Private Places*.

PLATE 57: Approaching Hubbard's Cave on the Dan Ryan Expressway, one of the nation's deadliest, most congested, and most famous Interstate highways, 1983.

PLATE 58: View north from the Canalport Street exit on the Dan Ryan, 1982.

PLATE 59: Motorcyclist/Chicago skyline, as seen from the Dan Ryan, 1983.

PLATE 60: Sears Tower/Moonrise on the Dan Ryan, 1984.

PLATE 61: Factory fire, along the Dan Ryan, 2 a.m., Canalport Street, 1985.

PLATE 62: Yellow kayak beneath the Dan Ryan, near the Chicago River, 1981.

PLATE 63: A homeless man's home beneath the Dan Ryan, next to the Chicago River, 1983.

PLATE 64: Winter garden, Wicker Park, 2004.

PLATE 65: Man reading by a fenced front garden, River West, 2000.

PLATE 66: Back garden, Lakeview, 2000.

PLATE 67: Side garden with yellow door, Wicker Park, 2000.

PLATE 68: Mini-garden, Wicker Park, 2002.

PLATE 69: English garden, Ravenswood, 2002.

PLATE 70: Head in garden, Pilsen, 2003.

Gallery VI

THE SUBURBS, 1992–2004

PHOTOGRAPHS BY BOB THALL AND TERRY EVANS

BOB THALL'S photographs in Gallery VI of *Chicago Portfolio* (Plates 71–78) come from *The New American Village*. His biographical note appears in Gallery I.

TERRY EVANS (b. 1944, Kansas City, Missouri) has taught photography at Columbia College Chicago for many years. She has received fellowships in photography from the John Simon Guggenheim Memorial Foundation and Mid-America Arts Alliance/National Endowment for the Arts. She is the author of *Revealing Chicago: An Aerial Portrait* (2005), *Disarming the Prairie* (1998), *The Inhabited Prairie* (1998), and *Prairie: Images of Ground and Sky* (1980). Her photographs are in the permanent collections of the Art Institute of Chicago, Baltimore Museum of Art, George Eastman House, Museum of Fine Arts in Houston, Museum of Modern Art, National Museum of American Art of the Smithsonian Institution, San Francisco Museum of Modern Art, and Whitney Museum of American Art, among others. Terry Evans's photographs in *Chicago Portfolio* (Plates 79–84) come from *Revealing Chicago*.

PLATE 71: Hoffman Estates, Cook County, 1993.

PLATE 72: Corporate complex at Schaumburg, Cook County, 1991.

PLATE 73: Higgins Road, Schaumburg, Cook County, 1996.

PLATE 74: Rooftop space and boulevard entrance, Ameritech building, Rolling Meadows, Cook County, 1992.

PLATE 75: Shopping at Schaumburg, Cook County, 1995.

PLATE 76: Living in Schaumburg, Cook County, 1995.

PLATE 77: Model home, Elk Grove Village, Cook County, 1995.

PLATE 78: The last barn, State Highway 59, Naperville, DuPage County, 1992.

PLATE 79: View of downtown Chicago from Schaumburg, Cook County, June 22, 2004.

PLATE 80: Housing subdivision in Buffalo Grove, Cook County, May 21, 2003.

PLATE 81: Former farm, Green Garden Country Club, Frankfort, Will County, September 23, 2003.

PLATE 82: Public swimming pool, Highland Park, Lake County, June 23, 2003.

PLATE 83: Backyard pools, Frankfort Square, Will County, September 17, 2003.

PLATE 84: Prairie Crossing, Graylake, Lake County, August 24, 2003. Nationally recognized as a model conservation community, the development has two commuter rail stations: one connecting it to O'Hare Airport and the other to downtown Chicago.

REFERENCES AND SELECTED READINGS

Berry, Brian J. L. *Chicago: Transformations of an Urban System*. Cambridge, MA: Ballinger Publications, 1976.

Berry, Brian J. L., and Frank E. Horton, eds. *Geographic Perspectives on Urban Systems*. Englewood Cliffs, NJ: Prentice-Hall, Inc., 1970.

Brown, James A., and Patricia J. O'Brien, eds. *At the Edge of Prehistory: Huber Phase Archaeology in the Chicago Area*. Kampsville, IL: Center for American Archaeology, 1990.

Burnham, Daniel H., and Edward H. Bennett. *Plan of Chicago*. Chicago: The Commercial Club, 1909.

Center Books on Chicago and Environs (see pages 253–54).

Changnon, Stanley A., Jr. "Evidence of Urban and Lake Influence on Precipitation in the Chicago Area." *Journal of Applied Meteorology* 19 (1980): 1137–59.

Chicago Department of Development and Planning. *Historic City: The Settlement of Chicago*. Chicago: City of Chicago, 1976.

Chrzastowki, Michael J., and Todd A. Thompson, "Late Wisconsinan and Holocene Coastal Evolution of the Southern Shore of Lake Michigan." In *Quaternary Coasts of the United States: Marine and Lacustrine Systems*, ed. Charles H. Fletcher III and John F. Wehmiller. Tulsa: Society for Sedimentary Geology (1992): 397-413.

Clark, Colin. "Urban Population Densities." *Journal of the Royal Statistical Society, Series A* 114 (1951): 490-96.

Clark, William A. V. *The California Cauldron: Immigration and the Fortunes of Local Communities*. New York: Guilford Press, 1998.

Colten, Craig E. *Industrial Wastes in the Calumet Area, 1869-1970: An Historical Geography*. Champaign, IL: Hazardous Waste Research and Information Center, State Water Survey Division, Illinois Department of Energy and Natural Resources, 1985.

Condit, Carl W. *Chicago, 1910-29: Building, Planning, and Urban Technology*. Chicago: University of Chicago Press, 1973.

Conzen, Michael P. "The Changing Character of Metropolitan Chicago." *Journal of Geography* 85 (1986): 224-36.

_____, ed. *Geographical Excursions in the Chicago Region*. Washington, D.C.: Association of American Geographers, 1995.

Cowles, Henry C., "The Physiographic Ecology of Chicago and Vicinity: A Study of the Origin, Development, and Classification of Plant Societies." *Botanical Gazette* 31 (1901), nos. 2 and 3.

Cronon, William. *Nature's Metropolis: Chicago and the Great West*. New York: W. W. Norton & Co., 1991.

Cutler, Irving. *Chicago: Metropolis of the Mid-Continent*. Chicago: Geographic Society of Chicago, 1973.

_____. *Chicago: Metropolis of the Mid-Continent*., 2nd ed. Dubuque: Kendall-Hunt Publishing Co., 1976.

_____, ed. *The Chicago Metropolitan Area: Selected Geographic Readings*. New York: Simon & Schuster, 1970.

Davis, James L. *The Elevated System and the Growth of Northern Chicago*. Studies in Geography, No. 10. Evanston: Northwestern University, Department of Geography, 1965.

Dear, Michael J., ed. *From Chicago to L.A.: Making Sense of Urban Theory*. Thousand Oaks, CA: Sage Publications, 2002.

DeVisé, Pierre. *Chicago's People, Jobs, and Homes: The Human Geography of the City and Metro Area*. 2 vols. Chicago: DePaul University, Department of Geography, 1964.

_____. "The Suburbanization of Jobs and Minority Employment." *Economic Geography* 52 (1976): 348-62.

Drake, St. Clair, and Horace Cayton. *Black Metropolis*. New York: Harcourt, Brace and Co., 1945.

Evans, Terry. Text by Charles Wheelan. *Revealing Chicago: An Aerial Portrait*. New York: Harry N. Abrams, in collaboration with Openlands Project and Metropolis 2020, 2005.

Grese, Robert. *Jens Jensen: Maker of Natural Parks and Gardens*. Baltimore: Johns Hopkins University Press, in association with the Center for American Places, 1992.

Grossman, James R. *Land of Hope: Chicago, Black Southerners, and the Great Migration*. Chicago: University of Chicago Press, 1989.

Grossman, James R., et al., eds. *Encyclopedia of Chicago*. Chicago: University of Chicago Press, 2004.

Harris, Chauncy D., and Edward L. Ullman. "The Nature of Cities." *Annals of the American Academy of Political and Social Science* 242 (1945): 7-17.

Hoyt, Homer. *One Hundred Years of Land Values in Chicago: The Relationship of the Growth of Chicago to the Rise in its Land Values, 1830-1933*. Chicago: University of Chicago Press, 1933.

Jablonsky, Thomas J. *Pride in the Jungle: Community and Everyday Life in Back of the Yards Chicago*. Baltimore: Johns Hopkins University Press, in association with the Center for American Places, 1993.

Local Community Fact Book, Chicago Metropolitan Area, 1980. Chicago: Chicago Fact Book Consortium, University of Illinois at Chicago, 1984.

Mayer, Harold M. "Localization of Railway Facilities in Metropolitan Centers as Typified by Chicago." *Journal of Land and Public Utility Economics* 20 (1944), 299-315.

Mayer, Harold M., and Richard C. Wade. *Chicago: Growth of a Metropolis*. Chicago: University of Chicago Press, 1969.

Meyer, Alfred H. "The Kankakee Marsh of Northern Indiana and Illinois." *Papers of the Michigan Academy of Science, Arts and Letters* 21 (1935): 359-96.

Pacyga, Dominic A., and Ellen Skerrett. *Chicago, City of Neighborhoods*. Chicago: Loyola University Press, 1986.

Park, Robert E., and Ernest W. Burgess, eds. *The City*. Chicago: University of Chicago Press, 1925

Petersen, Jon A. *The Birth of City Planning in the United states, 1840–1917*. Baltimore: Johns Hopkins University Press, in association with the Center for American Places, 2003.

Pierce, Bessie Louise. *A History of Chicago*. Vol. I, *The Beginning of a City, 1673-1848*. Chicago: University of Chicago Press, 1937.

_____. *A History of Chicago*. Vol. II, *From Town to City, 1848-1871*. New York: Alfred A. Knopf, 1940.

Reinemann, Martin W. "The Pattern and Distribution of Manufacturing in the Chicago Area." *Economic Geography* 36 (1960): 139-44.

Rose, Harold. *The Black Ghetto: A Spatial Behavioral Perspective*. New York: McGraw-Hill, 1971.

Scott, Allen J., and Edward W. Soja, eds. *The City: Los Angeles and Urban Theory at the End of the Twentieth Century*. Berkeley: University of California Press, 1996.

Shearmur, Richard, and Mathieu Charron, "From Chicago to L.A. and Back Again: A Chicago-Inspired Quantitative Analysis of Income Distribution in Montreal." *The Professional Geographer* 56 (2004): 108-26.

Soja, Edward W. *Postmetropolis: Critical Studies of Cities and Regions*. Malden, MA: Blackwell Publishers, 2000.

Solzman, David M. *Waterway Industrial Sites: A Chicago Case Study*. Research Paper No. 107. Chicago: University of Chicago, Department of Geography, 1967.

Taira, Atsushi, "Spatial Characteristics and Strategies of Japanese-Affiliated Companies in the Midwest of the United States: Localization or Specialization?" *Geographical Review of Japan* 75 (2002): 730-49.

Thall, Bob. *The New American Village*. Baltimore: Johns Hopkins University Press, in association with the Center for American Places, 1999.

_____, with an essay by Peter Bacon Hales. *The Perfect City*. Baltimore: Johns Hopkins University Press, in association with the Center for American Places, 1994.

Wade, Louise Carroll. *Chicago's Pride: The Stockyards, Packingtown and Environs in the Nineteenth Century*. Urbana: University of Illinois Press, 1987.

Whyte, William H., Jr. *The Organization Man*. New York: Simon and Schuster, Inc., 1956.

Wille, Lois. *Forever Open, Clear and Free: The Struggle for Chicago's Lakefront*. Chicago: Henry Regnery Co., 1972.

Willman, H. B. *Summary of the Geology of the Chicago Area*. Illinois State Geological Survey, Circular 460. Urbana: State of Illinois, Department of Registration and Education, 1971.

Wilson, William H. *The City Beautiful Movement*. Baltimore: Johns Hopkins University Press, in asssociation with the Center for American Places, *1989*.

Zorbaugh, Harvey. *The Gold Coast and the Slum*. Chicago: University of Chicago Press, 1929.

CENTER BOOKS ON CHICAGO AND ENVIRONS
2001–2006

About the Series

The mission of this series is to enhance the public's understanding of, appreciation for, and affection for Chicago's natural and built environment. The series and its first ten volumes were supported in part by a generous grant from the Graham Foundation for Advanced Studies in the Fine Arts of Chicago. The University of Chicago Press has served as the distributor and marketer of the series and as the publisher of Volume II and co-publisher of Volume X. The following list presents the first eleven books published in the series during its first five years. It began with the momentous release of Volume I at a public ceremony co-hosted by the City of Chicago, Richard M. Daley, Mayor, the Chicago Park District, and the Parkways Foundation just two days after 9/11.

— GEORGE F. THOMPSON, *series founder and director*

VOLUME I   Bachrach, Julia Sniderman, with a foreword by Bill Kurtis. *The City in a Garden: A Photographic History of Chicago's Parks*. Santa Fe, NM, and Harrisonburg, VA: Center for American Places, in association with the Chicago Park District, 2001.

VOLUME II   Greenberg, Joel. *A Natural History of the Chicago Region*. Chicago and London: University of Chicago Press, in association with the Center for American Places, 2002.

VOLUME III   Thall, Bob, with an afterword by Ross Miller. *City Spaces: Photographs of Chicago Alleys*. Santa Fe, NM, and Harrisonburg, VA: Center for American Places, 2002.

VOLUME IV   Crane, Barbara, with a foreword by Michael A. Weinstein and
               an essay by John B. Rohrbach. *Urban Anomalies: Chicago*. Chicago:
               Barbara Crane Educational Trust, in association with the Center for
               American Places, 2002.

VOLUME V   Wolke, Jay, with a conclusion by Dominic A. Pacyga. *Along the
               Divide: Photographs of the Dan Ryan Expressway*. Santa Fe, NM,
               and Staunton, VA: Center for American Places, in association
               with Columbia College Chicago, 2004.

VOLUME VI   Stochl, Gary, with an introduction by Bob Thall. *On City Streets:
               Chicago, 1964–2004*. Santa Fe, NM, and Staunton, VA: Center for
               American Places, in association with Columbia College Chicago, 2004.

VOLUME VII   Temkin, Brad, with an introduction by Rod Slemmons.
               Private Places: Photographs of Chicago Gardens. Santa Fe,
               NM, and Staunton, VA: Center for American Places, in
               association with Columbia College Chicago, 2005.

VOLUME VIII   Fortino, Scott, with an introduction by Judith Russi Kirshner.
               *Institutional: Photographs of Jails, Schools, and Other Chicago
               Buildings*. Santa Fe, NM, and Staunton, VA: Center for
               American Places, in association with Columbia College Chicago, 2005.

VOLUME IX   Thall, Bob. *At City's Edge: Photographs of the Chicago Lakefront*.
               Santa Fe, NM, and Staunton, VA: Center for American Places, in
               association with Columbia College Chicago, 2005.

VOLUME X   Hudson, John C. *Chicago: A Geography of the City and Its Region*.
               Chicago and London: University of Chicago Press and Santa Fe and
               Staunton: Center for American Places, 2006.

VOLUME XI   Frederking, William, with an afterword by Brandy Savarese.
               At Home. Santa Fe and Staunton: Center for American Places,
               in association with Columbia College Chicago, 2006.

INDEX

ABOUT THE AUTHOR

John C. Hudson (b. 1941) is Professor of Geography at Northwestern University in Evanston, Illinois, where he also directs the Environmental Sciences Program. He is a native of Milton, Wisconsin, and completed his B.S. in geography at the University of Wisconsin-Madison. After earning his M.A. and Ph.D. degrees in geography at the University of Iowa, he taught briefly at several universities before joining the Northwestern faculty in 1971. He has been a John Simon Guggenheim Memorial Foundation Fellow, and he served as editor of the *Annals of the Association of American Geographers* from 1976–1981. He is the author of *Geographical Diffusion Theory* (1972), *Plains Country Towns* (1985), which received the first John Brinkerhoff Jackson Prize from the Association of American Geographers in 1985, *Crossing the Heartland: Chicago to Denver* (1992), *Making the Corn Belt: A Geographical History of Middle-Western Agriculture* (1994), and *Across This Land: A Regional Geography of the United States and Canada* (2002).

The Center for American Places is a tax-exempt 501(c)(3) nonprofit organization, founded in 1990, whose educational mission is to enhance the public's understanding of, appreciation for, and affection for the natural and built environment. Underpinning this mission is the belief that books provide an indispensable foundation for comprehending and caring for the places where we live, work, and explore. Books live. Books endure. Books make a difference. Books are gifts to civilization.

With offices in Santa Fe, New Mexico, and Staunton, Virginia, Center editors bring to publication as many as thirty books per year under the Center's own imprint or in association with publishing partners such as the University of Chicago Press. The Center is also engaged in other outreach programs that emphasize the interpretation of place through art, literature, scholarship, exhibitions, and field research. The Center's Cotton Mather Library in Arthur, Nebraska, its Martha A. Strawn Photographic Library in Davidson, North Carolina, and a ten-acre reserve along the Santa Fe River in Florida are available as retreats upon request. The Center is also affiliated with the Rocky Mountain Land Library in Colorado.

The Center strives every day to make a difference through books, research, and education. For more information, please send inquiries to P.O. Box 23225, Santa Fe, NM 87502, U.S.A., or visit the Center's Website: www.americanplaces.org.

ABOUT THE BOOK:
The text for *Chicago: A Geography of the City and Its Region* was set in Granjon. The paper for the text is Thai A, 120 gsm weight, and for the portfolio Chinese Goldeast, 157 gsm weight. The book was printed and bound in China.

FOR THE CENTER FOR AMERICAN PLACES:
George F. Thompson, President and Publisher
Amber K. Lautigar, Publishing Liaison and Associate Editor
Ashleigh A. Frank, Editorial and Production Assistant
Brian M. Venne, Chelsea Miller Goin Intern
Kristine K. Harmon, Manuscript Editor
David Skolkin, Book Designer and Art Director
Dave Keck, of Global Ink, Inc., Production Coordinator